The Best
AMERICAN
SCIENCE &
NATURE
WRITING
2022

The Best AMERICAN SCIENCE & NATURE WRITING™ 2022

Edited and with an Introduction
by AYANA ELIZABETH JOHNSON

JAIME GREEN, *Series Editor*

MARINER BOOKS
New York   Boston

FIRST EDITION

ISSN 1530-1508
ISBN 978-0-358-61529-3

22 23 24 25 26   LSC   10 9 8 7 6 5 4 3 2 1

Contents

Humans Are a Part of Nature

Ways of Knowing

Futures We Could Have

Foreword

I HAVE BEEN THINKING a lot about beauty.

For the last two years I've worked at a desk set in front of a window. Behind my computer screen, past the words I'm reading or writing, past the faces of people (sorry) on video calls, I look, at least in summer, into an almost solid wall of green. The lawn stretches, hemmed in by hedges, to the oak and maple and tree of heaven at the back of the yard, reaching higher than I can see through my window. In winter, the leaves thin, and I see the houses on the other side of a small creek. But in summer, which it is as I'm writing this, that verdant wall is all I want to see.

This has been the view my eyes rested on while I thought about my own writing, which is mostly about space. Planets, stars, the possibility of life. Nothing so familiar as all this green, instead either cold and black or barren rock, or unimaginably alien. But I've realized that the trees and birds and vast cosmos are all the same thing to write about. The whole world is.

The pieces drawn together in this collection illuminate some of that web. And I'm grateful for it, because there isn't always a view out a window; it isn't always summer. The sweep of a spiral galaxy draws our eyes and attention much more than the black of space between the galaxies ever could. But there's so much beauty to understand in the things we can't see.

There's beauty in the things we think unbeautiful, too. In "The Body's Most Embarrassing Organ Is an Evolutionary Marvel," Katherine J. Wu applies her literary talents to the humble anus, shunned by polite society, subject to an intense giggle factor, but also, Wu writes, "shrouded in scientific intrigue." Anuses are important. Their origins are unknown and debated, holding clues to many mysteries of evolution. "In the beginning, there was nothing," Wu writes with stark simplicity. And then, somehow, there were anuses. Many of them familiar, but others quite strange. "One unusually aerated specimen, a type of polyclad flatworm, sports multiple anuses that speckle its backside like feces-spewing freckles." It's fascinating, and it's also poetry. The image of a backside speckled with feces-spewing freckles might not be beautiful, but the facts are fascinating, and "feces-spewing freckles" might be a new "cellar door."

Lacy M. Johnson similarly sanctifies a subject that, I will be honest, makes me cringe in photos: slime molds. Like that flatworm butt, she manages to describe, with delicate artistry on the page, a scene that I'm sure would make me retch if I saw it in life: "The dampness has darkened the flower bed, and from the black mulch has emerged what looks like a pile of snotty scrambled eggs in a shade of shocking, bilious yellow. As if someone sneezed on their way to the front door, but what came out was mustard and marshmallow."

But these are not just exercises in aesthetics. Johnson uses slime molds as the entry to a meditation on interconnectedness. Slime molds challenge our notions of categorization and hierarchy, single-celled blobs that nonetheless "can solve mazes, learn patterns, keep time, and pass down the wisdom of generations." They defy, according to Johnson, our taxonomies and systems of value. "In high school I learned that humans reigned over five kingdoms: animals, plants, fungi, protists, and bacteria. We came only from ourselves; we owed one another nothing." But slime molds, able to meld with others of their kind and split into spores, show that individuality is not the only way, no fact of nature.

I used to think there were a few different kinds of science-and-nature writing. There was the beautiful, poetic kind, there was the news kind, and there was the political kind; writing that stirred the senses, writing that informed, and writing that called for action. Lyrical writing invited us to peer at tiny things and gape at the impossibly large, evoking wonder and delight; informative writing gave us important information, no nonsense, just the state of the world; activist writing highlighted a problem, wrought by humans, and showed the ways we were fucking up the world, evoking anger, indication, sorrow, and, hopefully, change.

The fact is, there is no distinction here at all. Human action has brought us to a place where every new way we can find to love the world is a call to action. And every revelation about the ways we've marked the climate with our presence is proof of deep connection. And a call to act.

Jeff Goodell reminds us of this—the need to act, as well as the power of polemic—in "Our Summer from Hell." He uses facts as well as the strong rhythm of language to craft an authoritative demand that we move past climate crisis wake-up calls to actual action. He cites the progress of the last several years, a paragraph of accomplishments and back-pats. Then: "This is all good. This is all important. But if this is what it means to 'wake up' to the risks of the climate crisis, then we truly are fucked."

Yet Goodell doesn't stump for doom. He offers anger and frustration, but also a loving lamentation for a simpler world lost, a world that never was: "The big problem America faces here in the early years of the twenty-first century is that we built our world with the idea that we live on a stable, steady planet. The land is here, the ocean is there, and forever it shall be. The rains will come, but they will be rains like we always knew it to rain. It will get hot, but no hotter than it ever has."

Jessica Plumb also offers another way to see the changing world, the tension between human space and wilderness, in "A River Reawakened": "The beauty and diversity of Olympic National Park draws visitors from around the world. Many come to experience its 'ancient groves' of old-growth forest. To me, this familiar landscape does not feel ancient. I imagine this is what the world was like when it was young. Freshly scrubbed by glaciers, with terrain like a restless teenager, prone to earthquakes and pulsing with life." The world has never been static, never still. It is just that right now, we are the most forceful agents of change.

We can see one major facet of this change through a trio of pieces in this anthology that seem to me to form a song in three parts. In "Humanity Is Flushing Away One of Life's Essential Elements," Julia Rosen traces the history of humanity's use of this crucial element as a fertilizer, once cycled through food and human waste back to the soil, now threatened by the disruption of that chain. In "Why Combining Farms and Solar Panels Could

Transform How We Produce Both Food and Energy," Chris Malloy investigates a twinned path forward for solar energy and agriculture, sketching a future where sheep graze in the shadows of solar panels, where solar panels' shade gives thirsty crops a cooler place to grow. And in "A Recipe for Fighting Climate Change and Feeding the World," Sarah Kaplan writes of a quest to change how the planet eats. Since the advent of agriculture, she writes, "Three annual grasses—wheat, rice, and corn—became the foundation of human diets and human civilization." But scientists developing a perennial grain hope not only to make it easier to feed future generations, but to heal ecosystems that monoculture farming has harmed, slowing erosion, stowing carbon, and reducing the need for fertilizers. "Is it a natural evolution from the past 10,000 years of annual agriculture?" Kaplan writes. "Or something more like a midcourse correction?"

All of the stories in this book ask these questions, in one way or another. We can no longer assume, as Goodell writes, that "[t]he land is here, the ocean is there, and forever it shall be." Our future is not the one of the ancient, young Earth. It is one made by human hands. With all the scars that entails—marks of harm, and marks of healing.

The view out my office window is nothing special by the measure of most natural views. I live in a common suburb, my little bit of earth two minutes from Target and the highway; it's only from the narrow view of my desk, and the beneficence of summertime trees, that I can get a view without any other houses. Maybe I only treasure this view because I spent the previous years of my life in a city, sneaking glimpses of sky between the buildings, treasuring every rare afternoon spent in a park. But it's not that I wasn't "in nature" then, of course. Not because I could walk to a park or see the sky, and not because of pigeons and tenacious weeds. But because it's all nature, it's all things we've done with wood and stone; we're all flesh and bone and the electric impulses of cognition. Same as every animal, same roots as every tree. Everything we've done to the world is actually something we've done with it. As Johnson reminds us in her piece about slime molds, "Any system that claims to impose a hierarchy of value on this web is, like petri dishes and toasters and even the very idea of nature, a human invention. Superiority is not an inherent reality of the natural world."

The green view out my window and the pieces in this book remind me that to love the world is to care for it, that to see its beauty is to protect its survival, and that nature is not just nature. The world is not "out there." I hope this book does the same for you as well.

Nominations for *The Best American Science and Nature Writing 2023* are open. For more information or to nominate work, please visit jaimegreen.net/BASN.

JAIME GREEN

Introduction

IN A MELTING, flaming, unhealthy world, how do we hold on to our biophilia, make sense of the mess, and navigate toward solutions? This is a big question. Each essay in this collection is a piece of the answer.

This anthology attempts to go beyond compendium and toward discourse—interweaving ideas and putting them into conversation with one another through simple proximity. While each piece stands resoundingly on its own, here, grouped into themes, each finds a place in an arc that offers a directionality from observations to actions.

As I began combing through hundreds of articles to curate this collection, a handful of themes emerged among the pieces I found most enthralling. These themes then guided the rest of my selections among the plethora of wonderful writings published last year:

1. NATURE IS MAGNIFICENT—Given the overwhelming intensity of the world these last few years, I was tempted to fill the entire book with delightful essays about the wonders of biodiversity. The evolution of the anus, the intelligence of slime, the growth of galaxies, the circulatory systems of giraffes, the far-flungness of animal migrations. (I mean, what's not to love?)

2. NATURE IS ROILED—Nature's magnificence is in turmoil. Disasters are no longer natural, groundwater is rising, California is parched, forests are on fire, the ocean is loud, and plastic is everywhere. No sugarcoating here, but lots of sense-making.

3. HUMANS ARE A PART OF NATURE—We are but one of eight million or so species on this planet. From foraging for food to our relationships with bears, from our role in the phosphorus cycle to fighting against pollution and poisons, from climate-fueled migrations to what to do with our bodies after we die, these pieces elucidate our place in the big picture.

4. WAYS OF KNOWING—Being human in rapidly changing societies and on a rapidly changing planet is complex. From Indigenous knowledge and practices to artificial intelligence and physics, to how we use language, how we understand each thread, and how they braid together.

5. FUTURES WE COULD HAVE—Where do we go from here? What do more gentle ways of living on this planet look like, from how we feed our bodies and our appetites for electricity to where we work to who is in charge? Also, a shout-out to beavers. This section has the greatest number of pieces because—goodness, do we need writing that shows us ways forward out of our intertwined crises of climate, economy, inequality, and biodiversity.

Within each section the essays were ordered deliberately so that specific ideas would intermingle and flow one into the next. However, while reading, certainly feel free to skip around if a particular title or section catches your attention—life is short, eat dessert first!

At the most basic level, this book is filled with what I, personally, find poignant and fascinating. It is populated through the lenses of my predilections and biases. I am biased toward surprises and solutions. I am biased toward diversity in an expansive sense of that term—therein lies the stability of ecosystems and the magic of humanity. I am biased toward planet Earth and toward ecology, and away from the rest of the universe and technology. I am also (nerd alert!) biased toward spreadsheets and built one as I worked through a heap of articles to put together a cohesive selection.

What's that? You'd like to hear more about this spreadsheet? So glad I rhetorically asked myself this question on your behalf. There were columns for title, author, publication, and word count. And a row for each article. One cool thing about this "Best of . . ." series is that writers and editors can (via a simple online form) submit articles for consideration, so that's what initially populated the spreadsheet. This is seemingly a superegalitarian process that resulted in 800+ submissions, except that I expect many younger writers or writers from smaller publications may not know this opportunity exists, or for various reasons (shout-out to social dynamics and imposter syndrome) some writers may be unlikely to nominate themselves or be nominated. Thus, to initially understand the diversity of the pool I was selecting from, I added columns for race, gender, age (a rough guess by decade), location where the writer is based, and location (if any) the piece focused on. At that point it became clear that this was on track to be an exceedingly white and coastal collection of pieces from highbrow publications. In other words, it would have been a narrower "best of" than I wanted to present. So, I expanded the scope of the search.

In the lingo of the game show *Who Wants to Be a Millionaire?*, it was time to "phone a friend." I began asking my science-y, nature-y, well-read colleagues for suggestions of their favorite writers, publications, and articles. I scoured the internet. I thought back to all the pieces I had recently read that had sparked my curiosity and augmented my understandings. In addition to newspaper and magazine articles, I considered climate newsletters like Emily Atkin's Heated and *Atmos* magazine's *The Frontline,* publications that I've turned to repeatedly in the last few years as I tried to make sense of things. I read and read. I added many rows to the spreadsheet.

With this diversified list, I considered what would be most valuable to share with you. Of course, so much more excellent writing was published last year than could be contained between these covers. (I *highly* recommend checking out the list of honorable mentions at the end of the book.) So how did I narrow the purview and choose? The first answer is that I was guided by the five emergent themes described above. The second answer is that I begged the publisher for both more time and to include more essays than in previous editions. (Thank you for granting my wishes!) And third, I wanted to introduce you to some wonderful new writers whose work you may not yet have encountered.

Perhaps the simplest way to explain the curation approach I ended up using is by defining each word of the book's title.

"Best" is about elegant writing, yes, but also about important writing. You'll find some pieces in here that are less lyrical and more topically critical. But all of them are pieces that roused in me the immediate need to talk about them, to share the new (or new to me) information and insights they contained. "Best" includes deeply reported pieces that unfurl slowly over thousands of words, and pieces that were written on deadline with a constrained word limit and get straight to the point. Often getting an assignment for a larger piece is a privilege not afforded to younger journalists, journalists of color, or journalists at smaller or more newsy publications. So, to avoid excluding such writers, I didn't limit "best" to "epic." In lieu of a detailed rating system, I relied on whether I gasped or laughed or said "wow" while reading each of these pieces, and whether I rushed to share them with a "did you know?!" Or "you gotta read this!"

"American," as defined by this book's publisher, means the writing appeared in an American or Canadian publication. Canada, I'm sorry your name doesn't get on the cover, but know that I adore you.

"Science and nature" I interpreted both more expansively and more narrowly than might be anticipated. Another bias of mine is being more enamored with ecology, evolution, and anthropology than with technology, medicine, and engineering. In short, I lean toward the "nature" in "science and nature." And given that we are human, that our species is a supremely disruptive force on this planet, and that that disruption is manifesting in ways that horrifyingly exacerbate existing inequalities, that gets significant attention, too.

And "2022," well, all these pieces were actually published in 2021. Takes a little time to curate and publish these things. Lag time. Reading time. Thinking time. "Best" also has a *timeliness* to it, as in a piece's ability to capture and contextualize this moment in history. In a significant sense, this anthology is a time capsule.

Using these themes and interpretations, I selected the thirty-three pieces presented here (always love a palindrome). One-third of them were selected from the official submissions, and two-thirds from my own additional digging. Heartfelt thanks to Jaime Green, the series editor, for providing a short list of pieces for me to start with. And to Jessica Vestuto and the team at HarperCollins, thank you for entrusting me with this project.

I share all of this about the process of creating this collection because "best of" lists reflect the minds and hearts and worldviews of the list makers, a fact that is too often glossed over.

All this is to say, to be a curator is a serious (and joyful!) responsibility. It's a chance to point toward compelling and insightful writing and writers. It's a chance to elevate contributions that might otherwise go unrecognized. There is power in the anointing of "bestness" in a notable series such as this one, and I tried to honor that.

And about that joy. In a year where I, like many, have been recovering from burnout and saying "no" more often than ever, this project was an immediate and enthusiastic "yes!" I grew up with these "best of" books of short stories around the house, thanks to my English teacher mother. In 2020, with attention span shot and nerves frayed, I found myself turning to them for nuggets of wisdom and humanity. I hope that amid these pages you will find stories and facts and phrases that sing to you, as they did to me.

AYANA ELIZABETH JOHNSON

The Best
AMERICAN
SCIENCE &
NATURE
WRITING
2022

Nature Is Magnificent

KATHERINE J. WU

The Body's Most Embarrassing Organ Is an Evolutionary Marvel

FROM *The Atlantic*

TO PEER INTO THE SOUL of a sea cucumber, don't look to its face; it doesn't have one. Gently turn that blobby body around, and gaze deep into its marvelous, multifunctional anus.

The sea cucumber's posterior is so much more than an exit hole for digestive waste. It is also a makeshift mouth that gobbles up bits of algae; a faux lung, latticed with tubes that exchange gas with the surrounding water; and a weapon that, in the presence of danger, can launch a sticky, stringy web of internal organs to entangle predators. It can even, on occasion, be a home for shimmering pearlfish, which wriggle inside the bum when it billows open to breathe. It would not be inaccurate to describe a sea cucumber as an extraordinary anus that just so happens to have a body around it. As Rebecca Helm, a jellyfish biologist at the University of North Carolina at Asheville, told me, "It is just a really great butt."

But the sea cucumber's anus does not receive the recognition it deserves. "The moment you say 'anus,' you can hear a pin drop in the room," Helm said. Bodily taboos have turned anuses across the tree of life into cultural underdogs, and scientific ones, too: not many researchers vocally count themselves among the world's anus enthusiasts, which, according to the proud few, creates a bit of a blind spot—one that keeps us from understanding a fundamental aspect of our own biology.

The appearance of the anus was momentous in animal evolution, turning a one-hole digestive sac into an open-ended tunnel. Creatures with an anus could physically segregate the acts of eating and defecating, reducing the risk of sullying a snack with scat; they no longer had to finish processing one meal before ingesting another, allowing their tubelike body to harvest more energy and balloon in size. Nowadays, anuses take many forms. Several animals, such as the sea cucumber, have morphed their out-hole into a Swiss Army knife of versatility; others thought that gastrointestinal back doors were so nice, they sprouted them at least twice. "There's been a lot of evolutionary freedom to play around with that part of the body plan," Armita Manafzadeh, a vertebrate morphology expert at Brown University, told me.

But anuses are also shrouded in scientific intrigue, and a fair bit of squabbling. Researchers still hotly debate how and when exactly the anus first arose, and the number of times the orifice was acquired or lost across different species. To tap into our origins, we'll need to take a squarer look at our ends.

In the beginning, there was nothing. The back ends of our animal ancestors that swam the seas hundreds of millions of years ago were blank, relegating the entry and exit of all foodstuffs to a single, multipurpose hole. Evolutionary echoes of these lifeforms still exist in corals, sea anemone, jellyfish, and a legion of marine worms whose digestive tract takes the form of a loose sac. These animals are serially monogamous with their meals, taking food in one glob at a time, then expelling the scraps through the same hole. (Contrary to what you might have read, not everyone poops.) These creatures' guts operate much like parking lots, subject to strict vacancy quotas that restrict the flow of traffic.

The emergence of a back door transformed those parking lots into highways—the linear "through-guts" that dominate body plans today. Suddenly, animals had the luxury of downing multiple meals without needing to fuss with disposal in between; digestive tracts lengthened and regionalized, partitioning into chambers that could extract different nutrients and host their own communities of microbes. The compartmentalization made it easier for animals to get more out of their meals, Andreas Hejnol, a developmental biologist at the University of Bergen, in Norway, told me. With the lengthening and uncorking of the end of the gut, he said, many creatures grew into longer and larger body forms, and started to move in new ways. (It would take several more eons for true buttocks—the fleshy, fatty accoutrements that flank the anuses of some animals, such as humans—to evolve. Some researchers I talked with are comfortable using *butt* to mean any anal or anus-adjacent structure; others are purists and consider the term strict shorthand for *buttocks* and *buttocks* alone.)

The benefits of bottoming out the gut are clear; how the back door was excavated isn't. Soft, squishy, bone-free holes aren't exactly fixtures of the fossil record, making just about *any* anus-heritage theory tough to prove. One of the oldest hypotheses holds that the anus and the mouth originated from the same solo opening, which elongated, then caved in at the center and split itself in two. The newly formed anus then moseyed to the animal's posterior. Claus Nielsen, a developmental biologist at the Natural History Museum of Denmark, is a fan of this theory. It's both reasonably parsimonious and evolutionarily equitable: In this scenario, neither the mouth nor the anus technically arose first; they emerged as perfect developmental twins.

Hejnol and others favor a different idea, in which the mouth formally preceded the anus, which spontaneously burst through the other end of the body. "It's a secondary breakthrough," Hejnol said. "The gut forms, then [makes] a connection to the outer world." Punching an extra hole in the body is not so difficult: Some worms have managed the feat dozens of times over. One unusually aerated specimen, a type of polyclad flatworm, sports multiple anuses that speckle its backside like feces-spewing freckles. Two others, a pair of sponge parasites called *Syllis ramosa* and *Ramisyllis multicaudata,* will twine their body through host tissues like a tapestry of tree roots, with each tip terminating in its own proprietary butthole; they have hundreds, perhaps thousands, in total. (It's not totally clear why these animals and others spawned an embarrassment of anuses, but in at least some cases, Hejnol thinks it's a logical outcome of a branched digestive system, which can more easily transport nutrients to a body's every nook and cranny.)

Hejnol and his colleagues are still amassing support for their hypothesis, but he said there's already some argument against the hole-splitting idea: animals don't generally express the same genes around their mouth and anus, a knock against the notion that the two openings are cut from the same developmental cloth. A better backstory for the orifice, he said, might involve a body plan stolen from the reproductive tract, which already naturally terminates at the animalian posterior.

If that theory pans out, though, it won't necessarily close the case on the anus's evolutionary start. A cursory glance at the animal tree of life might at first suggest that anal openings appeared about 550 million years ago, around the time our own bloblike ancestors straightened out into tubes. But Hejnol and many others think that the anus was so useful that animals independently evolved it at least half a dozen times, perhaps many more, and not necessarily in the same way. This timeline has other snags: Some creatures have since lost their anal opening—and some might have made theirs even further back in history.

One of the largest potential wrinkles in the smooth anus narrative takes the shape of a comb jelly—a gelatinous animal that vaguely resembles a translucent Darth Vader helmet and is thought to be at least seven hundred million years old. As far back as the 1800s, scientists have been puzzling over comb jellies' back end, and whether they were excreting formal feces from a set of strange-looking pores. More than a century passed before their acts of defecation were finally caught on camera, by the biologist William Browne of the University of Miami and his colleagues, who filmed one of the amorphous creatures taking a big fishy dump in the lab. When the clip debuted at a 2016 conference, "everyone in the hall audibly gasped," Helm, who attended the lecture, told me. If comb jellies were pooping, that poop had to be coming out of some sort of rear hole. Perhaps, some said, the history of the anus ran far deeper in time than many had thought.

In the months after Browne's team published its findings, scientists sparred repeatedly over their significance. Some hailed the discovery as revolutionary. But others, Hejnol among them, argued that the now-infamous video didn't signify all that much dogmatic change, and may not be hard to reconcile with what's long been known. Comb jellies probably cooked up their anuses independently of other animals and happened upon a similar blueprint by chance; there's no telling when exactly that might have occurred. Such a scenario would leave the chronology of our own anus, which emerged out of a different line of creatures at a separate point in time, intact.

The various possibilities aren't easy to prove or disprove. Just as new apertures can rupture into being, useless ones can disappear, as seems to have been the case with brittle stars and mites, which stitched their ancestral anuses shut. Some ambivalent creatures might even gouge out *transient* anuses—holes that come and go on an as-needed basis. (A 2019 study by the biologist Sidney Tamm suggested that some comb-jelly anuses could fall into this category of "sometimes-butts," as Manafzadeh calls them.)

Many of the animals that have managed to keep some version of the anus embellished upon it, and now harbor an organ of immense extravagance. Turtles, like sea cucumbers, breathe through their butt. Young dragonflies suck water into theirs, then spew it out to propel themselves forward. Scorpions jettison their posterior when attacked from behind, evading capture but tragically losing their ability to poop (and eventually dying with their abdomen full of excrement). Lacewing larvae incapacitate termite prey with the toxic flatulence they emit from their end—"they literally KO their enemies with death farts," Ainsley Seago, an entomologist at the Carnegie Museum of Natural History, told me.

Some of the most intriguing (and NSFW) back ends are all-purpose anus analogues called cloacae, which merge the terminal parts of the digestive, urinary, and reproductive tracts into a single opening—essentially an evacuation foyer for outbound feces, urine, eggs, and sperm. Cloacae are fixtures among birds, reptiles, and amphibians, and although they tend to get a bad rap, their internal architecture is actually quite sophisticated, Patricia Brennan, a cloaca expert at Mount Holyoke College, in Massachusetts, told me. They can also be quite convenient: When female birds mate with unsatisfactory males, they can simply eject the subpar sperm and begin the process anew. Cloacae have been around for so long, Hejnol added, that they could even represent the evolutionary bridge between the reproductive and digestive tracts that helped lead to some of the first anuses.

Still, cloacae come with risks: "You have all your digestive waste pretty much in direct contact with genitalia," basically a gnarly infection just waiting to happen, Brennan said. Any live young who pass through the reproductive tract could also be imperiled by the proximity to poop-borne pathogens. Perhaps that's why human anuses ventured off on their own.

*

Whatever the reason behind it, the partitioning that did away with the cloaca made human anuses, as Manafzadeh said, "completely boring." As far as exit holes go, ours are standard-issue, capable of little more than extruding waste from the gut, with no frills to speak of.

The only redeeming quality of humans' humdrum posterior hole is the feature we evolved to cushion it: our infamous buttocks, the most voluminous ones documented to date, thanks to our bizarre tendency to strut around on our two primate legs. "Our bipedalism is obligate; it's special; it's the only way we get around," Darcy Shapiro, an anthropologist, told me. That pattern of locomotion reshaped the pelvis, which in turn reoriented our muscles. The gluteus maximus—the hefty muscle that powers our ability to run and climb—swelled in lockstep and blanketed itself in a cozy layer of fat that some scientists think serves as an energy reserve. Anuses aside, "our buttocks are the real innovation here," Manafzadeh said.

Evolution blew the human butt out of proportion; our cultural norms quickly followed suit. We regard one another's bums with lust, disgust, and guilty fascination. We shrink them, we sculpt them; we sexualize them. We rap about them with abandon. They, in return, make it much easier to sprint, but much harder to keep our rear ends clean. Our anus is a sheep dressed in very fabulous wolf's clothing, and we simply cannot deal.

Maybe that's part of why humans are so often embarrassed by their posteriors, and, by extension, so many others. We even opt for *butt* as a euphemism for *anus* in casual conversation. Buttocks aren't anuses, but they do cloak them, physically and perhaps figuratively. They obscure the idea that, from its very start, our digestive end has been a wonder. It cracked open our ancestors' evolutionary path and made our own existence possible. Maybe it's time we made like a pearlfish and got comfortable with what's between those cheeks.

LACY M. JOHNSON

What Slime Knows

FROM *Orion*

Nothing from nothing ever yet was born.
  —Lucretius, "On the Nature of Things"

IT IS SPRING IN HOUSTON, which means that each day the temperature rises and so does the humidity. The bricks of my house sweat. In my yard the damp air condenses on the leaves of the crepe myrtle tree; a shower falls from the branches with the slightest breeze. The dampness has darkened the flower bed, and from the black mulch has emerged what looks like a pile of snotty scrambled eggs in a shade of shocking, bilious yellow. As if someone sneezed on their way to the front door, but what came out was mustard and marshmallow.

I recognize this curious specimen as the aethalial state of *Fuligo septica,* more commonly known as "dog vomit slime mold." Despite its name, it's not actually a mold—not any type of fungus at all—but rather a myxomycete (pronounced MIX-oh-my-seat), a small, understudied class of creatures that occasionally appear in yards and gardens as strange, Technicolor blobs. Like fungi, myxomycetes begin their lives as spores, but when a myxomycete spore germinates and cracks open, a microscopic amoeba slithers out. The amoeba bends and extends one edge of its cell to pull itself along, occasionally consuming bacteria and yeast and algae, occasionally dividing to clone and multiply itself. If saturated with water, the amoeba can grow a kind of tail that whips around to propel itself; on dry land the tail retracts and disappears. When the amoeba encounters another amoeba with whom it is genetically compatible, the two fuse, joining chromosomes and nuclei, and the newly fused nucleus begins dividing and redividing as the creature oozes along the forest floor, or on the underside of decaying logs, or between damp leaves, hunting its microscopic prey, drawing each morsel inside its gooey plasmodium, growing ever larger, until at the end of its life, it transforms into an aethalia, a "fruiting body" that might be spongelike in some species, or like a hardened calcium deposit in others, or, as with *Stemonitis axifera,* grows into hundreds of delicate rust-colored stalks. As it transitions into this irreversible state, the normally unicellular myxomycete divides itself into countless spores, which it releases to be carried elsewhere by the wind, and if conditions are favorable, some of them will germinate and the cycle will begin again.

From a taxonomical perspective, the *Fuligo septica* currently "fruiting" in my front yard belongs to the Physaraceae family, among the order of Physarales, in class Myxogastria, a taxonomic group that contains fewer than a thousand individual species. These creatures exist on every continent and almost everywhere people have looked for them: from Antarctica, where *Calomyxa metallica* forms iridescent beads, to the Sonoran Desert, where *Didymium eremophilum* clings to the skeletons of decaying saguaro cacti; from high in the Spanish Pyrenees, where *Collaria chionophila* fruit in the receding edge of melting snowbanks, to the forests of Singapore, where the aethalia of *Arcyria denudata* gather on the bark of decaying wood, like tufts of fresh cotton candy.

Although many species are intensely colored—orange, coral pink, or red—others are white or clear. Some take on the color of what they eat: ingesting algae will cause a few slime molds to turn a nauseous green. *Physarum polycephalum,* which recently made its debut at the Paris Zoo, is a bright, egg yolk yellow, has 720 sexual configurations and a vaguely fruity smell, and appears to be motivated by, among other things, a passionate love of oatmeal.

Throughout their lives, myxomycetes only ever exist as a single cell, inside which the cytoplasm always flows—out to its extremities, back to the center. When it encounters something it likes, such as oatmeal, the cytoplasm pulsates more quickly. If it finds something it dislikes, like salt, quinine, bright light, cold, or caffeine, it pulsates more slowly and moves its cytoplasm away (though it can choose to overcome these preferences if it means survival). In one remarkable study published in *Science,* Japanese researchers created a model of the Tokyo metropolitan area using oat flakes to represent population centers and found that *Physarum polyceph-alum* configured itself into a near replica of the famously intuitive Tokyo rail system. In another experiment, scientists blasted a specimen with cold air at regular intervals, and found that it learned to expect the blast, and would retract in anticipation. It can solve mazes in pursuit of a single oat flake, and later, can recall the path it took to reach it. More remarkable still, a slime mold can grow indefinitely in its plasmodial stage. As long as it has an adequate food supply and is comfortable in its environment, it doesn't age and it doesn't die.

Here in this little patch of mulch in my yard is a creature that begins life as a microscopic amoeba and ends it as a vibrant splotch that produces spores, and for all the time in between, it is a single cell that can grow as large as a bath mat, has no brain, no sense of sight or smell, but can solve mazes, learn patterns, keep time, and pass down the wisdom of generations.

How do you classify a creature such as this? In the ninth century, Chinese scholar Twang Ching-Shih referred to a pale-yellow substance that grows in damp, shady conditions as *kwei hi*, literally "demon droppings." In European folklore, slime mold is depicted as the work of witches, trolls, and demons—a curse sent from a neighbor to spoil the butter and milk. In Carl Linnaeus's *Species Plantarum*—a book that aspired to list every species of plant known at the time (nearly seven thousand by the 1753 edition)—he names only seven species of slime molds. Among those seven we recognize *Fuligo* in the species he calls *Mucor septicus* ("rotting mucus"), which he classifies, incorrectly, as a type of fungus.

At the time, life hadn't been studied in detail at the microscopic level, and Linnaeus's taxonomic classifications, few of which have withstood the scrutiny of modern science, were based almost entirely on observable phenotype—essentially, how they looked to the naked eye. He placed *Mucor septicus* in the same genus as *Mucor mucedo*, because, well, they both looked like mucus. The fruiting bodies of both of these species looked like a type of fungus, and fungus looked like a type of plant.

We now call Linnaeus the "father" of taxonomy. Though he wasn't the first to try to impose order on nature—naturalists, philosophers, and artists had constructed their own schema as far back as Aristotle—he was the first to classify our own species within his system, naming us *Homo sapiens* and placing us, scandalously, within the animal kingdom. That idea, that humans were "natural" beings, "Anthropomorpha" in the same order as chimpanzees and gorillas and sloths, drew the ire of Linnaeus's fellow naturalists, whose intellectual lineage could be traced back at least to Aristotle, who had ordered the physical world along a continuum from inanimate objects through plants and then to animals. These "ladders" or "scales of ascent," in turn, inspired the "Great Chain of Being"—the Christian worldview, central to European thought from the end of the Roman Empire through the Middle Ages, that ordered all of creation from lowest to highest, beginning with the inanimate world, through plants and animals, placing humans just below angels, and angels just below God. If anything like slime mold appeared there, it would no doubt be near the very bottom, just above dirt.

Over time, Linnaeus revised his classifications of Homo sapiens, naming "varieties" that at first corresponded to what he saw as the four geographic corners of the planet, but which became hierarchical, assigning different intellectual and moral value based on phenotypes and physical attributes. The idea that humans could and should be ordered—that some were superior to others, that this superiority had a physical as well as social component—was deeply embedded in many previous schemata. But Linnaeus's taxonomy, unlike the systems that came before, gave these prejudices the appearance of objectivity, of being backed by scientific proof. When Darwin's *On the Origin of Species* was published in 1859, it was on the foundation of this "science," which had taught white Europeans to reject the idea of evolution unless it crowned them in glory.

But the history of taxonomic classification has always been about establishing hierarchy, beginning with Linnaeus, who offered the world his binomial naming system as well as its first three taxonomic kingdoms: plants (Regnum Vegetabile), animals (Regnum Animale), and minerals (Regnum Lapideum, which Linnaeus himself later abandoned). Ernst Haeckel—biologist, artist, philosopher, and fervent disciple of Darwin—expanded Linnaeus's model in 1866. To the plant and animal kingdoms, Haeckel added a third: Protista, for the various microscopic organisms known but not understood at that time. These included sponges and radiolaria and myxomycetes, the term Heinrich Friedrich Link had proposed for slime molds in 1833. Developments in microscope technology in the nineteenth century had given Haeckel and his fellow biologists a glimpse into the world of organisms too small to see with the naked eye, and with it, a keen interest in accounting for the evolutionary relationships of all species on Earth in ever more minute detail. Haeckel called this new science phylogeny, and he filled pages and pages of his works with intricately illustrated phylogenetic trees—beautiful in their execution but diabolical in their implications. In perhaps his best-known illustration, "The Pedigree of Man," he places "man" at the highest point of a great oak, while apes, ungulates, "skull-less animals," worms, and amoeba are lower down because he saw them as less evolved and therefore closer to the root of creation. Elsewhere he similarly categorized humanity into as many as twelve different *species* with different evolutionary histories—white Europeans, in his view, being the most evolved, important, and civilized.

Taxonomy has evolved in the centuries since Haeckel and Linnaeus, but much of their thinking still remains. Even if science no longer views humans as divided into different and unequal species, we continue to refer to "race" as if it were a natural, biological category rather than a social one created in service of white supremacy. The myth that humans are superior to all other species—that we are complex and intelligent in a way that matters, while the intelligence and complexity of other species does not—also exists in service to white supremacy, conferring on far too many people an imagined right of total dominion over one another and the natural world.

In high school I learned that humans reigned over five kingdoms: animals, plants, fungi, protists, and bacteria. We came only from ourselves; we owed one another nothing. I learned this in my parents' church, too, that the world was made for men, that every life (my own included) was under their dominion. I did not learn until college about a taxonomic category that superseded kingdom, proposed in the 1970s by biologists Carl Woese and George Fox and based on genetic sequencing, that divided life into three domains: Bacteria, Eukarya, and Archaea, a recently discovered single-celled organism that has survived in geysers and swamps and hydrothermal vents at the bottom of the ocean for billions of years.

Perhaps a limit of our so-called intelligence is that we cannot fathom ourselves in the context of time at this scale, and that so many of us fail, so consistently, to marvel at any lives but our own.

I remember a recent visit to the Morian Hall of Paleontology at the Houston Museum of Natural Science. I moved with the exhibit through geologic time, beginning with trilobite fossils from more than five hundred million years ago, toward creatures that become larger and more terrifying before each of five extinction events, in all of which climate change has been a factor. Each time, millions of species have disappeared from the planet, but thanks to small, simple organisms, life has somehow carried on.

The hall's high ceilings and gentle lighting make it feel more like a contemporary art exhibit than a scientific display, and though scientists might object to this approach, for a layperson like me, it fostered wonder, and wonder has often been my antidote to despair. At the very end of the winding geologic maze, I encountered mammals and megafauna before arriving in the smallest exhibit in the entire hall, where a wall case contained the fossilized skulls of the various human lineages, mapping the web of their links and connections. So much damage has been done by the lie that this world belongs only to a few, that some lives matter more than others. The consequences of that lie have changed Earth more in a few decades than in the previous several million years. Outside, the next extinction looms.

But it is also possible to move through the exhibit in the opposite way, beginning with the urgency of the present and journeying back through time—to pass through doorways in this history that show us unexpected connections, to see the web of life spread out before us in all its astonishing diversity. Any system that claims to impose a hierarchy of value on this web is, like petri dishes and toasters and even the very idea of nature, a human invention. Superiority is not an inherent reality of the natural world.

Humans have been lumbering around the planet for only a half million years, the only species young and arrogant enough to name ourselves *sapiens* in genus *Homo*. We share a common ancestor with gorillas and whales and sea squirts, marine invertebrates that swim freely in their larval phase before attaching to rocks or shells and later eating their own brain. The kingdom Animalia, in which we reside, is an offshoot of the domain Eukarya, which includes every life-form on Earth with a nucleus—humans and sea squirts, fungi, plants, and slime molds that are ancient by comparison with us—and all these relations occupy the slenderest tendril of a vast and astonishing web that pulsates all around us and beyond our comprehension.

The most recent taxonomies—those based on genetic evidence that evolution is not a single lineage, but multiple lineages, not a branch that culminates in a species at its distant tip, but a network of convergences—have moved away from their histories as trees and chains and ladders. Instead, they now look more like sprawling, networked webs that trace the many points of relation back to ever more ancient origins, beyond our knowledge or capacity for knowing, in pursuit of the "universal ancestors," lifeforms that came before metabolism, before self-replication—the several-billion-year-old plasmodial blobs from which all life on Earth evolved. We haven't found evidence for them yet, but we know what we're looking for: they would be simple, small, and strange.

A few years ago, near a rural village in Myanmar, miners came across a piece of amber containing a fossilized *Stemonitis* slime mold dating from the mid-Cretaceous period. Scientists were thrilled by the discovery, because few slime mold fossils exist, and noted that the one-hundred-million-year-old *Stemonitis* looks indistinguishable from the one oozing around forests today. Perhaps slime mold hasn't evolved much in that time, they speculated. Recent genetic analyses have suggested that slime molds are perhaps as old as one or two billion years—which would make them hundreds of millions of years older than plants and would mean they pulled themselves out of the ocean on their cellbows at a time when the only land species were giant mats of bacteria. One special ability of slime molds that supports this possibility is their capacity for cryptobiosis: the process of exchanging all the water in one's body for sugars, allowing a creature to enter a kind of stasis for weeks, months, years, centuries, perhaps even for millennia. Slime molds can enter stasis at any stage in their life cycle—as an amoeba, as a plasmodium, as a spore—whenever their environment or the climate does not suit their preferences or needs. The only other species who have this ability are the so-called "living fossils" such as tardigrades and Notostraca (commonly known as water bears and tadpole shrimp, respectively). The ability to become dormant until conditions are more favorable for life might be one of the reasons slime mold has survived as long as it has, through dozens of geologic periods, countless ice ages, and the extinction events that have repeatedly wiped out nearly all life on Earth.

Slime mold might not have evolved much in the past two billion years, but it has learned a few things during that time. In laboratory environments, researchers have cut *Physarum polycephalum* into pieces and found that it can fuse back together within two minutes. Or each piece can go off and live its separate life, learn new things, and return later to fuse together, and in the fusing, each individual can teach the other what it knows, and can learn from it in return.

Though, in truth, "individual" is not the right word to use here, because "individuality"—a concept so central to so many humans' identities—doesn't apply to the slime mold worldview. A single cell might look to us like a coherent whole, but that cell can divide itself into countless spores, creating countless possible cycles of amoeba to plasmodium to aethalia, which in turn will divide and repeat the cycle again. It can choose to "fruit" or not, to reproduce sexually or asexually or not at all, challenging every traditional concept of "species," the most basic and fundamental unit of our flawed and imprecise understanding of the biological world. As a consequence, we have no way of knowing whether slime molds, as a broad class of beings, are stable or whether climate change threatens their survival, as it does our own. Without a way to count their population as a species, we can't measure whether they are endangered or thriving. Should individuals that produce similar fruiting bodies be considered a species? What if two separate slime molds do not mate but share genetic material? The very idea of separateness seems antithetical to slime mold existence. It has so much to teach us.

In 1973, in a suburb of Dallas, a sudden, particularly spectacular appearance of *Fuligo septica* across lawns sparked a panic. Firemen blasted the plasmodia with water, breaking the creatures to pieces, but those pieces continued to slime around and grow larger. The townspeople speculated that an indestructible alien species had invaded Earth, perhaps recalling the plot of the 1958 movie *The Blob* starring a young Steve McQueen. Scientists arrived in the panicked neighborhood to take samples, reassuring the community that what they had experienced was just a stage in the life cycle of a poorly understood organism: "a common worldwide occurrence," they said. "Texas scientists think backyard blob is dead," read a headline in the *New York Times*.

The slime mold in my yard is also dead, I think. The aethalia is pale, hardened, and calcified, with the texture and color of a summer cast protecting a child's broken arm, browned by a season without washing. A breath of wind arrives and black dust lifts from the slime mold's surface, blown toward the edge of my yard, and the next one over. And the next.

Spring in Houston is the season for working in the garden. We replant our tall ornamental grasses, killed in the recent unseasonable freeze; ours are a hybrid *Pennisetum* species, from the family Poaceae, a large taxonomic group that also contains *Zea* (a genus that includes corn), *Oryza* (rice), *Saccharum* (sugar cane), and *Triticum* (wheat). Fungi live on and among these plants, bringing them water and nourishment through the threadlike mycelium to keep them alive and aiding their decomposition when they die. As the plants decompose, they provide the food that bacteria eat, and myxomycete amoeba prey on these bacteria when they hatch from their spores. We plunge our shovels and hands in the dirt, the living substrate—alive in ways I have only just begun to fathom. We plant the grasses, fill the holes, lay down fresh mulch. We collect our tools and retreat indoors to the comforts of our home—our refrigerated food, our instant oatmeal, our beloved air-conditioning.

Days later, I am leaving my house to walk the dogs, the air hanging dankly all around me, and out of the corner of my eye I see dozens of bright coral pink beads scattered across the surface of the fresh mulch, a new species I learn is *Lycogala epidendrum,* "wolf's milk slime mold." I know very little about it but receive this marvelous arrival in the only way I know how: we are made by, and for, one another.

ARIANNA S. LONG

Too Big for the Universe

FROM *Scientific American*

JUST LIKE TREES, people, and stars, galaxies have life cycles. A galaxy is born when enough gas and stars coalesce to form a coherent structure—perhaps it starts as one cloud of gas and slowly gathers mass, or maybe it builds up from the collision of two or more clouds. Either way, once formed, a galaxy spends its lifetime making stars, using its reservoirs of gas to create tiny furnaces where nuclear fusion burns elements to release light and energy. A galaxy deemed "alive" shines strongly in ultraviolet light, a signal of young, bright, and hot stars. As those stars age, their light changes from hot and blue to cool and yellow or red. When a galaxy contains mostly yellow and red stars and emits little to no ultraviolet light, we consider it retired, or "red-and-dead." Eventually, if massive enough, it becomes a spheroidal blob, known as an elliptical galaxy, that will likely never birth a new star again.

All around us in nearby space—within, say, three hundred million to six hundred million light-years—astronomers see dead or dying elliptical galaxies gathered in great ensembles called galaxy clusters. These clusters hold the fossilized remains of the most massive galaxies ever formed—hundreds to thousands of them slowly dancing around one another, gravitationally bound forever in their permanent graves.

But galaxy clusters present a problem for astronomers. Most clusters seem to have been established by the time the universe was only half of its current age. That means the galaxies within those clusters must have birthed most of the stars they contain early in cosmic history. It appears that these galaxies grew to the size of the Milky Way or larger but up to ten billion years more quickly. The young galaxy clusters, called protoclusters, where these galaxies formed must have been incredibly violent and active places, full of galaxies producing stars at a furious pace. Our current understanding of physics cannot quite explain how they could have grown so big so quickly.

Only recently have astronomers had the telescopic tools necessary to find protoclusters, which are very distant (their light often travels ten billion years or more to reach us) and frequently hide their most massive galaxy members behind dust. In the past few years, scientists have discovered two protoclusters that are providing an unprecedented window into cluster growth. Follow-up observations have revealed that they are, in fact, active and huge—so huge that they challenge our understanding of galaxy formation. If we can solve the riddle posed by galaxy clusters, we may redefine our understanding of the evolution of the universe.

Hunting for Starbursts

The most common type of star-forming galaxy produces roughly one to tens of suns' worth of stars a year. These are often called normal star-forming galaxies. The Milky Way is in this class. Normal star-forming galaxies are metaphorical tortoises, forming stars slowly and steadily over the course of ten billion years or so, remaining blue and disklike and depleting their reservoirs of fresh gas (fuel for new stars) at a leisurely pace.

Galaxies that produce hundreds to thousands of stars every year are known as starburst galaxies. These are the hares in galaxy evolution. In perhaps three hundred million years at most, these galaxies burst into existence, form as many stars as possible as quickly as possible and, in a cosmic blink of an eye, run out of fuel. Starbursts live fast and die young. Astronomers think that they are the best candidate ancestors for the massive, dead elliptical galaxies we see in clusters today.

It stands to reason that if we looked deep enough into space, we would find protoclusters filled with starburst galaxies—tomorrow's clusters of dead galaxies. Yet finding starburst galaxies in protoclusters has proved challenging. Until recently, most of our methods for spotting clusters were developed to preferentially select dying elliptical galaxies or the hot gas that pervades the space between them. Elliptical galaxies and hot intracluster gas appear at the later stages of galaxy cluster evolution, so we need new methods to find their bluer, more star-forming infant counterparts. To make things more difficult, protoclusters are often spread far apart on the sky because the galaxies have yet to fully coalesce into the dense structures we see today. When our most famous and precise telescopes have cameras that span only the width of a pencil (the Hubble Space Telescope, for instance), it is not surprising that we cannot piece together protocluster puzzle pieces that are spread across the sky at distances more than 100 times greater than our telescope's field of view.

Other methods of searching, such as systematically surveying large swaths of the sky, tend to miss starburst galaxies because they are often obscured by dust. The exceptional stellar growth in starbursts generates an overabundance of heavy metals that are produced in the explosive deaths of stars. Once dispersed into space, heavy elements such as iron, carbon, and gold collide to form complex dust molecules that absorb and obscure ultraviolet and optical light. Think of the reddening sun during wildfire season: dust dims hotter, more energetic blue light while letting redder light sneak through. The result is that starburst galaxies are nearly invisible when viewed with optical and ultraviolet telescopes, but they shine like beacons when viewed in the cooler infrared spectrum.

All of this means that until recently, the tools to find and study protoclusters usually missed a key population of galaxies. From the late 1990s through the early 2010s, the Submillimeter Common-User Bolometer Array, the Herschel Space Observatory, the South Pole Telescope, and the Spitzer Space Telescope revolutionized our understanding of the dust-obscured universe by unveiling millions of galaxies that were previously invisible. Starting about fifteen years ago, astronomers began studying the clustering properties of dusty starbursts, and they found that these powerhouses live preferentially near other large and actively star-forming galaxies. But the state of technology was still behind our ambitions; the resolution of infrared and millimeter telescopes was still so low that multiple galaxies would get blended into one large object, even if those galaxies were far apart but lay along the same line of sight. The age of the infrared universe was here, but we needed sharper and more sensitive instruments to fully comprehend what we were seeing.

Finally, in 2013, the Atacama Large Millimeter/submillimeter Array (ALMA) arrived. High in the Chilean desert, this collection of nearly seventy radio dishes works together as a single telescope, reaching resolutions up to six hundred times sharper than that of the Herschel telescope. ALMA has transformed many corners of astronomy, including galaxy evolution. (I know several people with tattoos dedicated to this telescope.) The observatory is excellent at detecting dusty, gaseous stellar nurseries throughout star-forming galaxies. With it, astronomers have discovered systems that are both shocking and exciting.

Surprising Behemoths

In 2018 two separate teams of astronomers used ALMA to study the brightest infrared objects they could find in the distant universe. Each team discovered a different conglomeration of dusty, starbursting galaxies that were previously blended together, hiding as one in surveys taken by the first generation of infrared telescopes. SPT2349-56, a group of fourteen galaxies, and the Distant Red Core (DRC), a group of ten galaxies, were both found growing and thriving in different corners of the universe, when the cosmos was only 10 percent of its current age. We see both these budding protoclusters undergoing extreme bursts of star formation—each group birthing nearly ten thousand times as much stellar mass a year as the Milky Way does—across volumes only half the size of our Local Group (which includes our own galaxy plus Andromeda and several smaller galaxies). Estimates of the protoclusters' individual gas reservoirs tell us that if these galaxies continued to form stars at such excessive rates, they would exhaust their fuel supply in just a few hundred million years and become the massive red-and-dead elliptical galaxies that are ubiquitous in fully grown clusters. Moreover, they would complete this cycle well before the present era.

The discovery of these two dust-obscured protocluster cores presented a promising new lens for studying cluster growth, but we were still missing an important part of the picture. The best way to "weigh" a galaxy is to measure the light from its adult star population, which requires data from across the electromagnetic spectrum. But until recently, all observations of protoclusters living in the first two billion years of the universe had been conducted within a narrow energy spectrum (in either the optical or the infrared). Then, in September 2018, my colleagues and I were able to observe, for the first time, ultraviolet and optical emission from a dusty, starbursting protocluster as seen twelve billion years ago: the Distant Red Core. Using the Hubble Space Telescope, the Gemini Observatory, and the Spitzer Space Telescope, we captured the multiwavelength perspective necessary to more deeply understand this structure's past and future.

Waiting for Hubble Space Telescope data can be anticlimactic. You know the day that your patch of the sky is scheduled for observation, but you have no idea when you will actually receive your data: you just have to wait for an email notification to tell you to check the archive. The day our protocluster observations were scheduled, I checked my email what felt like every two minutes. I was disappointed when it was time to go to bed and nothing had arrived in my in-box.

The following morning, against my partner's protests, I rolled out of bed immediately and went straight to my computer to see if the data had at last come in. Fortunately, it had been delivered a few hours past midnight. I commenced the download, dancing impatiently like a child waiting for her turn to unwrap presents. Finally, I opened the image. No words can describe how it feels to be the first person to glimpse a part of the universe that no one else has seen. I felt compelled to take a moment to inspect each star and galaxy in the field, to acknowledge their existence. Eventually I snapped back to myself and zoomed in on the part of the sky I was interested in. I saw something remarkable.

This little region of space is violent. At least half of the galaxies there were so messily shaped they must have recently crashed into other nearby galaxies or were still in the process of doing so. When we measured the population of adult stars in these galaxies, we found something incredible—so incredible that it may pose a problem for our current understanding of the universe. Already at this early era, some of the Distant Red Core galaxies had formed three times more stars than our own Milky Way has—but in just a fraction of the time. Yet simulations of the universe based on known physics struggle to produce galaxies this massive so early on. This incongruity between simulation and observation exacerbates a problem that we have known about since the discovery of dusty, star-forming galaxies. Modeling the extreme pace and density of star formation seen in starbursts is difficult because physics predicts the simulated galaxies should either shred themselves apart or heat up so much that they blow out all of the fresh gas needed to grow large enough to match what we see today.

The protocluster as a whole presents another problem: it is shockingly massive. When I first measured it, I could not believe my numbers. I knocked on doors in my department to make sure I was doing the calculations right. Two weeks later I brought the results to a conference to show my collaborators. One said, "There must be a bug in your code." Another asked, "Are you sure you're not double counting somewhere?" (Turns out there was a small bug in my code, but it was not enough to explain the huge measurement.) Eventually, after double-checking my calculations and trying out different methods, the measurements became undeniable. The Distant Red Core seems to be too big for our universe. We do not know how it could have gotten so large in such a short time.

To better understand its bulk and how much mass was in the form of stars, we zeroed in on the size of the dark matter halo around this protocluster. Dark matter is the most abundant form of matter in any given galaxy and in the universe as a whole. All galaxies and clusters are thought to be surrounded by blobs, or halos, of this mysterious stuff. And although it is invisible and poorly understood, dark matter leaves a clear gravitational signal. There are a variety of ways to infer the amount of dark matter in a given astronomical object, and to cover those methods would require an additional article (or five).

Suffice it to say, we weighed the dark matter component of the Distant Red Core, and according to our simulations, it contains nearly the largest allowable halo mass at that period in the history of the universe. This apparent overabundance of dark matter means that the DRC may be so large that it violates the laws of our universe as we understand them. When we fast-forward our simulations to estimate what the DRC may look like after evolving twelve billion years to the present day, we find that it may grow to be larger than the largest known galaxy cluster, El Gordo. Although we have a healthy margin of error on our dark matter calculation (meaning it could be overestimated), the discrepancy looks even worse when we consider the fact that our observations are capturing only a small percentage of the likely galaxy cluster members; there are probably more galaxies in the DRC that were simply out of the narrow field of view of our telescopes and thus not included in our calculations. This mismatch most likely will grow as we continue surveying and studying this protocluster.

Rethinking the Time Line

Our investigation of the Distant Red Core, along with the discoveries of other potentially similar protoclusters, forces us to reconsider our understanding of galaxy cluster formation. Because the galaxies in clusters are likely to be some of the first galaxies ever, we must determine how such massive objects could form so quickly. Doing so is not just an issue of constraining the physical mechanics and chemistry of star formation inside the first galaxies. It is also a matter of investigating the timing of the conditions that lead dark matter to gravitationally collapse into halos, seeding galaxies. Is it possible that galaxies and structure began forming earlier in the universe than we thought? What does that mean for our understanding of the formation of the first elements? Could these galaxies have forged the right ingredients to build stars with habitable planets around them—and perhaps hosted some of the first forms of life in the universe?

Some of these questions probably will not be answered during my lifetime, but I and other astrophysicists are working hard and fast to address the others. Already we are carrying out more observations of these known protoclusters across the electromagnetic spectrum. We are also developing new methods for identifying large samples of dusty protocluster candidates. With more examples, we may be able to determine whether protoclusters such as the Distant Red Core are examples of a common, yet previously invisible, phase of galaxy evolution that all clusters go through or just rarities. Observers and theorists are forming new collaborations to learn how early in the history of the universe conditions were right for protoclusters akin to those we have discovered—

pockets of space overdense with tremendous rates of star formation and outsize masses.

The best way to test our physical models is to look at extremes. In the next few years, these colossal congregations of exceptional galaxies will be putting humanity's grasp of the cosmos to the test.

BOB HOLMES

Heads Up! The Cardiovascular Secrets of Giraffes

FROM Knowable Magazine

TO MOST PEOPLE, giraffes are merely adorable, long-necked animals that rank near the top of a zoo visit or a photo-safari bucket list. But to a cardiovascular physiologist, there's even more to love. Giraffes, it turns out, have solved a problem that kills millions of people every year: high blood pressure. Their solutions, only partly understood by scientists so far, involve pressurized organs, altered heart rhythms, blood storage—and the biological equivalent of support stockings.

Giraffes have sky-high blood pressure because of their sky-high heads that, in adults, rise about six meters above the ground—a long, long way for a heart to pump blood against gravity. To have a blood pressure of 110/70 at the brain—about normal for a large mammal—giraffes need a blood pressure at the heart of about 220/180. It doesn't faze the giraffes, but a pressure like that would cause all sorts of problems for people, from heart failure to kidney failure to swollen ankles and legs.

In people, chronic high blood pressure causes a thickening of the heart muscles. The left ventricle of the heart becomes stiffer and less able to fill again after each stroke, leading to a disease known as diastolic heart failure, characterized by fatigue, shortness of breath, and reduced ability to exercise. This type of heart failure is responsible for nearly half of the 6.2 million heart failure cases in the United States today.

When cardiologist and evolutionary biologist Barbara Natterson-Horowitz of Harvard and UCLA examined giraffes' hearts, she and her students found that their left ventricles did get thicker, but without the stiffening, or fibrosis, that would occur in people. The researchers also found that giraffes have mutations in five genes related to fibrosis. In keeping with that find, other researchers who examined the giraffe genome in 2016 found several giraffe-specific gene variants related to cardiovascular development and maintenance of blood pressure and circulation. And in March 2021, another research group reported giraffe-specific variants in genes involved in fibrosis.

And the giraffe has another trick to avoid heart failure: the electrical rhythm of its heart differs from that of other mammals so that the ventricular-filling phase of the heartbeat is extended, Natterson-Horowitz found. (Neither of her studies has been published yet.) This allows the heart to pump more blood with each stroke, allowing a giraffe to run hard despite its thicker heart muscle. "All you have to do is look at a picture of a fleeing giraffe," Natterson-Horowitz says, "and you realize that the giraffe has solved the problem."

Natterson-Horowitz is now turning her attention to another problem that giraffes seem to have solved: high blood pressure during pregnancy, a condition known as preeclampsia. In people, this can lead to severe complications that include liver damage, kidney failure, and detachment of the placenta. Yet giraffes seem to fare just fine. Natterson-Horowitz and her team are hoping to study the placentas of pregnant giraffes to see if they have unique adaptations that allow this.

People who suffer from hypertension are also prone to annoying swelling in their legs and ankles because the high pressure forces water out of blood vessels and into the tissue. But you only have to look at the slender legs of a giraffe to know that they've solved that problem, too. "Why don't we see giraffes with swollen legs? How are they protected against the enormous pressure down there?" asks Christian Aalkjær, a cardiovascular physiologist at Aarhus University in Denmark who wrote about giraffes' adaptations to high blood pressure in the 2021 *Annual Review of Physiology*.

In part, at least, giraffes minimize swelling with the same trick that nurses use on their patients: support stockings. In people, these are tight, elastic leggings that compress the leg tissues and prevent fluid from accumulating. Giraffes accomplish the same thing with a tight wrapping of dense connective tissue. Aalkjær's team tested the effect of this by injecting a small amount of saline solution beneath the wrapping into the legs of four giraffes that had been anesthetized for other reasons. Successful injection required much more pressure in the lower leg than a comparable injection in the neck, the team found, indicating that the wrapping helped resist leakage.

Giraffes also have thick-walled arteries near their knees that might act as flow restrictors, Aalkjær and others have found. This could lower the blood pressure in the lower legs, much as a kink in a garden hose causes water pressure to drop beyond the kink. It remains unclear, however, whether giraffes open and close the arteries to regulate lower-leg pressure as needed. "It would be fun to imagine that when the giraffe is standing still out there, it's closing off that sphincter just beneath the knee," says Aalkjær. "But we don't know."

Aalkjær has one more question about these remarkable animals. When a giraffe raises its head after bending down for a drink, blood pressure to the brain should drop precipitately—a more severe version of the dizziness that many people experience when they stand up suddenly. Why don't giraffes faint?

At least part of the answer seems to be that giraffes can buffer these sudden changes in blood pressure. In anesthetized giraffes whose heads could be raised and lowered with ropes and pulleys, Aalkjær has found that blood pools in the big veins of the neck when the head is down. This stores more than a liter of blood, temporarily reducing the amount of blood returning to the heart. With less blood available, the heart generates less pressure with each beat while the head's down. As the head is raised again, the stored blood rushes suddenly back to the heart, which responds with a vigorous, high-pressure stroke that helps pump blood up to the brain.

It's not yet clear whether this is what happens in awake, freely moving animals, though Aalkjær's team has recently recorded blood pressure and flow from sensors implanted in free-moving giraffes and he hopes to have an answer soon.

So—can we learn medical lessons from giraffes? None of the insights have yet yielded a specific clinical therapy. But that doesn't mean they won't, says Natterson-Horowitz. Even though some of the adaptations are probably not relevant for hypertension in humans, they may help biomedical scientists think about the problem in new ways and find novel approaches to this far-too-common disease.

SONIA SHAH

How Far Does Wildlife Roam? Ask the "Internet of Animals"

FROM *The New York Times Magazine*

"I'M GOING TO DO A SET OF COOS," Calandra Stanley whispered into the radio. The Georgetown ornithologist and her team had been hunting cuckoos, in an oak-and-hickory forest on the edge of a Southern Illinois cornfield, for weeks. Droplets of yesterday's rain slid off the leaves above to those below in a steady drip. In the distance, bullfrogs croaked from a shallow lake, where locals go ice fishing in winter.

As dawn broke and the rising sun lit the top of the canopy, the cuckoo finally arrived to investigate. Within moments the bird was ensnared, squawking and thrashing and flapping his wings in a knot of black netting. Stanley slowly unfurled the net, cupping him in her hands. He had a slim handsome head, bright eyes, and long brown-and-white tail feathers soiled with a smear of feces. Stanley unceremoniously dumped him into a drawstring cloth bag and hooked it to a nearby tree. Inside the bag, he went silent, while the crew set up a tarp on a grassy opening nearby and spread out their gear.

With her instruments arrayed around her, Stanley gingerly drew the bird out of the bag, gripping him by his fuzzy white neck and scrawny legs. She blew all over his body, ruffling his down to look for the fat stores he might have built up for his coming journey. She clipped the claws at the end of his zygodactyl feet, two toes facing forward and two facing backward, and plucked one of his feathers, dropping it into a small manila envelope. She spread one of his wings so that she could get a blood sample. She measured him with calipers from various angles. He submitted, his eyes wide and glassy, except for when she took the width of his beak, which provoked a single, outraged yelp.

Then Stanley deposited a few drops of superglue to attach the object at the heart of her ministrations: a tiny solar-powered tracking device. She carried the cuckoo into a clearing a few feet away and asked me to open my palms, placing him inside them. Freed, he didn't hesitate for even a split second. As soon as she released her grip, he flew off into the trees, his feet ever so lightly grazing my open palm.

Last fall, teams of scientists began fanning out across the globe to stalk and capture thousands of other creatures—rhinos in South Africa, blackbirds in France, fruit bats in Zambia—in order to outfit them with an array of tracking devices that can run on solar energy and that weigh less than five grams. The data they collect will stream into an ambitious new project, two decades in the making and costing tens of millions of dollars, called the International Cooperation for Animal Research Using Space, or ICARUS, project. Each tag will collect data on its wearer's position, physiology, and microclimate, sending it to a receiver on the International Space Station, which will beam it back down to computers on the ground. This will allow scientists to track the collective movements of wild creatures roaming the planet in ways technically unimaginable until recently: continuously, over the course of their lifetimes and nearly anywhere on Earth they may go.

By doing so, ICARUS could fundamentally reshape the way we understand the role of mobility on our changing planet. The scale and meaning of animal movements has been underestimated for decades. Although we share the landscape with wild species, their movements are mostly obscure to us, glimpsed episodically if at all. They leave behind only faint physical traces—a few paw prints in the hardening mud of a jungle path, a quickly fading arc of displaced air in the sky, a dissipating ripple under the water's surface. But unlike, say, the sequence of the human genome, or the nature of black holes, where our fellow creatures go has not historically been regarded as a particularly pressing gap in scientific understanding. The assumption that animal movements are circumscribed and rare tended to limit scientific interest in the question. The eighteenth-century Swedish naturalist Carl Linnaeus, imagining nature as an expression of God's perfection, presumed each species belonged in its own singular locale, a notion embedded in his taxonomic system, which forms the foundation of a wide array of biological sciences to this day. Two centuries later, the zoologist Charles Elton, hailed as the "father of animal ecology," fixed species into place with his theory that each species nestles into its own peculiar "niche," like a pearl in a shell. Such concepts, like modern notions of "home ranges" and "territories," presumed an underlying stationariness in undisturbed ecosystems.

But over the last few decades, new evidence has emerged suggesting that animals move farther, more readily, and in more complex ways than previously imagined. And those movements, ecologists suspect, could be crucial to unraveling a wide range of ecological processes, including the spread of disease and species' adaptations to habitat loss. ICARUS will allow scientists to observe animal movements in near totality for the first time. It will help create what its founder, Martin Wikelski, a biologist at the University of Konstanz and managing director of the Max Planck Institute of Animal Behavior in Germany, calls the "internet of animals."

If successful, ICARUS will help us understand where animals go: the locations where they perish, the precise pathways of their migrations, their mysterious radiations into novel habitats— phenomena scientists have puzzled over for generations. "These are questions we've been trying to answer for thirty years," says the butterfly biologist Camille Parmesan, research director of the French National Center for Scientific Research. "It's fabulous." Peter Marra, an ecologist and the director of the Georgetown Environment Initiative at Georgetown University, agrees. ICARUS, he says, will be an "incredibly powerful tool to start asking these fundamental questions" in ecology, and to address "enormously vexing problems in conservation biology." The evolutionary ecologist Susanne Akesson, chairwoman of the Center for Animal Movement Research at Lund University in Sweden, notes that ICARUS "gives many possibilities for new research which has not been possible." The conservation ecologist Francesca Cagnacci, who coordinates a research consortium dedicated to studying the movement of terrestrial mammals, likens ICARUS to a sports car compared with a normal car. It will, she says, "take us to another level."

*

The ICARUS project challenges traditional paradigms whose tentacles run deep into science, politics, and culture. It isn't just that scientists were long unable to observe complex and long-distance wildlife movements, the way they had been unable to observe, say, the passage of DNA from parent to child. The scientific establishment presumed that what they couldn't see didn't exist. The absence of evidence of wild mobility, in other words, was taken as evidence of absence.

This wasn't a marginal notion with glancing significance. It was central to the way scientists, for decades, understood ecological processes, from climate change to how ecosystems established themselves and how diseases unfolded. When scientists predicted the impact of climate change, for example, many pictured immobile wild species marooned in newly inhospitable habitats, condemning them to extinction. When they considered the dispersal of seeds, which dictates the diversity and abundance of the plants that serve as the scaffolding of ecosystems, they dismissed the possibility that certain animals on the move played a role. Wild creatures like orchid bees, for example, could not possibly pollinate plants across long distances, scientists presumed, because they could not tolerate the heat stress of flying under direct sunlight; fruit-eating guácharos, or oilbirds, couldn't disperse seeds in the Venezuelan rainforest, because scientists thought the birds perched in their caves all day. The nineteenth-century naturalist Alexander von Humboldt dismissed the birds as parasites.

When scientists considered movements across barriers and borders, they characterized them as disruptive and outside the norm, even in the absence of direct evidence of either the movements themselves or the negative consequences they purportedly triggered. Popular hypotheses held that bats spread Ebola virus, for example, and gazelles foot-and-mouth disease. No one really knew where the bats or the gazelles went, though: The parallels between the intermittent and disruptive quality of epidemics and the presumed nature of wildlife movements spoke for themselves. Influential subdisciplines of biological inquiry focused on the negative impact of long-distance translocations of wild species, presuming that the most significant of these occurred not through the agency of animals on the move but when human trade and travel inadvertently deposited creatures into novel places. The result, experts in invasion biology and restoration biology said, could be so catastrophic for already-resident species that the interlopers should be repelled or, if already present, eradicated, even before they could cause any detectable damage.

Discoveries enabled by ICARUS, while impossible to predict, could have diffuse and wide-ranging implications. Findings that shed light on the factors that drive animal movement, for example, could help transform ecology from a field that traditionally describes the natural world and its inhabitants to one that can make predictions. Every year, billions of dollars depend on the ways in which wild species move and are distributed across the landscape, migrations that affect the abundance of fish we pull from the sea, the virulence of the pathogens we encounter, the predators that stalk our livestock, and the birds and flowers that grace our landscapes. But nobody knows precisely when the bats will arrive in any given forest, or why some butterflies shift into new ranges while others do not, or whether elephants that run shrieking in the forests have sensed an impending natural disaster, or why some martins return to their summer nests and others do not.

ICARUS could unlock that knowledge. It could enable scientists to unravel wild animals' social dynamics as they move around the globe in flocks, swarms, and colonies; to study what influence animals' conflicts and alliances with other species have on where they go and how they get there; and to chart the depth of their perceptions and the dynamism of their responses to the environmental phenomena they encounter on their journeys. Scientists may be able to detect shared strategies across populations, species, and taxa by observing the way various species navigate obstacles like roads and highways and the way they capitalize on environmental factors like currents in the sea and thermals in the air. Overlaying tracking data with data on weather, climate, and vegetation could reveal how the fragmentation of habitats affects animals' movement, which corridors they use to move, where they pause on their journeys, when they use environmental or atmospheric factors to facilitate their movement and how they might fare if those factors were to collapse or to change—drawing us closer to a future in which the movement of animals could be forecast, like the weather. The potential applications could include preventing outbreaks of disease that can precipitate pandemics, managing landscapes, and conserving biodiversity.

Almost certainly, prospectively tracking wild animals will reveal more extensive movements than previously known. A handful of tracking studies in recent years have established that wild animals wander across expansive ranges, oblivious to the boundaries of parks and conservation areas drawn to contain them. These studies uncovered several "megadispersals": a wolf that made it from Italy to France; a leopard that moved across three countries in southern Africa; mule deer that accomplished one of the longest land migrations of any species in North America. By tracking yellow-billed cuckoos, Stanley and Marra discovered that the birds move hundreds of kilometers, even on their breeding grounds, and are far less sedentary than previously thought. That finding torpedoes the traditional model of migration, in which the migratory journey is bracketed by stillness on both wintering and breeding grounds. ICARUS could mean a steady release of similarly confounding findings. It will "allow us to rewrite textbooks," Marra says.

Findings of novel long-distance peregrinations beyond the borders of recognized habitats unsettle deeply rooted ideas about our place in nature. They may suggest that wild animals have greater capacities for navigation and cognition than we've presumed, which could complicate the moral and political order we've justified on the basis of our supposedly unique cognitive abilities. They could suggest that we've misunderstood the role of geographic barriers in our migratory past and overestimated their role in the migrations to come. The planet may well be crisscrossed with "environmental highways" that usher wild migrants around the globe effortlessly, the way the trade winds ferried sailors across the Atlantic. Such a network has been proposed in modeling studies as an explanation for why migratory birds don't travel along the most direct paths but take looping, circuitous routes instead.

The delicate filigree of tracks that ICARUS exposes, in other words, could be "where the music is, where all the juice is," as Wikelski puts it. It's "the missing link that shapes everything."

Martin Wikelski is a soft-spoken fifty-five-year-old biologist with spiky dark hair and retro, black-framed glasses. We first spoke in 2017 and then reconnected over a series of video chats during the summer of 2020. A subtly mischievous expression animated his angular features as he told me, in the particular singsong lilt of some native German speakers' English, about his childhood dreams of knowing where animals go.

He remembers being about ten and peering into the abandoned swallows' nests in the eaves of his grandfather's barn in
Bavaria in winter, wondering why they had vanished. It was an
absence that had mystified European thinkers for centuries. The
sixteenth-century Swedish writer Olaus Magnus claimed that the
swallows spent their winters submerged in lakes; the English minister Charles Morton suggested that they flew to the moon. When a
teacher told Wikelski, in an offhand way, that the twenty-gram birds
flew thousands of miles away to Africa, it seemed to him an equally
fantastic tale.

But the methods available to confirm the swallows' itinerary—or
any other wild creatures'—were crude and few. To verify his teacher's pronouncement, Wikelski wrote a letter to relatives in South
Africa asking them if they'd seen any swallows there. He watched
a television program on bird banding and learned how to sneak
into the swallows' nests to affix tiny metal bands to the young birds
before they left, then traipsed around the half-dozen farmhouses
in the village to see if any returned to their vacated nests.

Fifteen years later, Wikelski had acquired a Ph.D. in zoology,
but wildlife tracking methods had only marginally improved. Commonly used "mark and recapture" techniques involved marking
individual animals in some way and then seeing if they could be
caught again, some distance away. Butterflies' wings might be in-
scribed with Magic Markers; birds' legs banded; or the landscape
itself wired with motion-sensing cameras to surreptitiously snap
photos of wild creatures as they skulked by. But such methods
could only corroborate that animals moved wherever scientists
thought to look for them. The marked birds and butterflies who
evaded recapture and the animals who strayed beyond the range of
motion-sensing cameras escaped scrutiny. Some scientists circum-
vented the confirmation bias of mark-and-recapture strategies by
outfitting animals with signal-emitting devices and then capturing
the signals on handheld or fixed receivers. But skeptics scoffed at
wildlife telemetry as a sterile substitute for the traditional fieldwork
of surreptitiously observing animals in the wild. At the time, wildlife
tracking was generally considered on "the margins of ecological
research," as Wikelski and colleagues would later write in a 2015
paper in *Science*. Attaching a tracking device to a wild animal generally required trapping it first, which was hard enough. On top

of that, the devices themselves could be expensive, awkward, and bulky, and capturing the signals often required scientists to embark on fruitless chases of their tagged subjects, receivers in tow.

Wikelski's first attempt to resolve the technical impasse unfolded in 2001 on Barro Colorado Island, a six-square-mile dripping jungle oasis in the middle of Gatun Lake in Panama, where he worked as a postdoctoral researcher for the Smithsonian Tropical Research Institute. The mammalogist Roland Kays, who would become a frequent collaborator, had been tracking nocturnal raccoonlike creatures called kinkajous nearby. To do it, he lured them into arboreal traps with bits of banana, then outfitted them with collars that emitted radio signals, which the thick vegetation readily absorbed. Then he spent his nights "trucking around the rainforest, chasing my kinkajous with my antenna," Kays recalls, "and thinking there must be some better way to do this."

The solution, Kays and Wikelski figured, was height. They devised a scheme to hoist receivers atop seven 130-foot towers dotted across the island. From their perch above the canopy, the receivers would be able to capture signals from tagged animals and automatically stream the data to a computer at the island's lab. They'd be able to track a range of species, simultaneously, across the entire island. They trapped and collared ocelots, sloths, and capuchins. They affixed transmitters to the bodies of orchid bees, using drops of superglue mixed with eyelash glue. They suffered the scratches of an upset anteater, then took turns dousing one another's wounds with alcohol. According to the scientific literature at the time, the island's watery borders marooned its residents, making the island "its own little universe in a way," Wikelski says. With a more comprehensive view of the animals' movements, they'd be able to answer questions about basic ecological functions, like how the movement of orchid bees and the ocelots' predation of rodents influenced the dispersal of seeds from trees and rare tropical plants.

But addressing such grand questions required that the scientists' subjects remain attached to their tags and within range of the island's receivers. They didn't. Wikelski and Kays discovered the tag from one of their ocelots at the bottom of the lake, scratched and hair-covered, presumably after passing through the body of a crocodile. At one point, the two scientists squeezed into the back of a helicopter to chase radio signals shimmering off the iridescent body of a tagged bee after it buzzed through the humid air across Gatun Lake.

It started to dawn on Wikelski that "all our preconceptions about this little universe are wrong," he told me. "Little bees fly off and on, so do toucans—pretty much everything that people said could not move around between places did." One evening, he and Kays were relaxing over cold drinks while overlooking the Panama Canal. They were joined by a retired radio engineer named George Swenson, who was among the first radio astronomers to track the Sputnik satellite that the Soviet Union secretly launched in 1957, by picking up the radio signal the satellite emitted. He went on to design and help build elaborate systems for scanning the heavens in search of other meaningful signals, including the National Radio Astronomy Observatory's array of more than two dozen radio telescopes in New Mexico that detect black holes.

The engineer was not impressed with the ecologists' 130-foot-high towers, Wikelski recalls. "You ecologists," Swenson said, "you're stupid. You have this big topic you could address, but you're thinking too small." The ecologists were like the early astronomers, studying disconnected slivers of the sky with their single telescopes. That hadn't allowed astronomers to understand the universe, which only became possible after they built arrays of telescopes to surveil all of space at once. To answer the big questions in ecology, Swenson suggested, ecologists had to track all the swimming, flying, and prowling creatures of the planet, everywhere, simultaneously. Hoisting receivers 130 feet in the air was not nearly high enough. The receiver had to be hundreds of miles away—in space.

Wikelski became "almost fanatical" about the idea, one of his colleagues at the Max Planck Society told a reporter for the scientific journal *Nature* in 2018. He spent months arranging a meeting at NASA to propose it. Their rejection did not deter him. He sought out new funding, new partners, new collaborators. According to the article, he became so preoccupied with getting the project off the ground that he nearly lost research funding for the Max Planck institute he directed. Wikelski's dogged pursuit of a lofty project like ICARUS most likely seemed as fanciful as trying to count all the leaves on a tree or the ripples in a lake.

*

The view that tracking wild mobility had limited value corresponded with a vision of the planet as fundamentally resistant to movement, littered with impassable obstructions like oceans, deserts, and mountains that constrained wild animals to their places. In mid-twentieth-century experiments that tried to characterize the physical challenge animals faced in migrating, for example, scientists trapped birds in wind tunnels—sealed tubes outfitted with fans that blew winds up to twenty mph steadily against them—and documented the birds' struggles to stay aloft. The wind tunnels simulated the conditions experts presumed flying creatures encountered in the wild: continuous, unrelenting resistance. Experiments like these concluded that long-distance migrations required herculean efforts, reinforcing presumptions about their peculiarity. According to the conventional wisdom, movement through even the most fluid mediums demanded propulsive force. As late as the 1940s, the roiling ocean was seen as a "place of eternal calm," as the biologist and writer Rachel Carson wrote, "its black recesses undisturbed by any movement of water more active than a slowly creeping current."

Skepticism about the prevalence of long-distance mass movements among wild species conformed, too, with the ways in which we negotiate settlement and migration in our own lives. Long-distance mass movements coordinated over short periods, in which hundreds of thousands of individuals left a certain place and then congregated again, weeks later, hundreds or thousands of miles away, required sophisticated coordination and navigation. Without the help of modern technology, Homo sapiens would not be able to achieve it as quickly as many wild species routinely do. Even with the help of advanced navigational technology and maps developed over generations, many of us get lost. That wild species—implicitly treated by many biologists and psychologists as "unthinking robots," as the zoologist Donald R. Griffin put it—might successfully accomplish superior feats of collective intelligence conflicted with the exceptionalism with which we made sense of ourselves in nature. As the ecologist Ran Nathan points out, "Many people consider animals very skillful, but not in cognition."

Over the decades that Wikelski struggled to launch ICARUS, technical advances in wildlife-tracking technology buoyed a newly emergent field of movement ecology, rattling norms about animal migration and helping to make the case for his project. The size and price of commercial GPS devices that could accurately pin-point geographic locations plummeted, from the early one-and-a-half-pound devices sold for thousands of dollars to fifty-dollar tags the size of a coin, allowing the boutique manufacturing firms that produce wildlife-tracking tags to churn out smaller, more accurate, and longer-lasting solar-powered tags. Wildlife telemetry entered what commentators called a "golden age," moving from the margins of ecological research toward the center. New, interdisciplinary research centers dedicated to the study of animal movement sprang up, including the CAnMove Center for Animal Movement Research at Lund University in Sweden, established in 2008, and the Minerva Center for Movement Ecology, which opened at the Hebrew University of Jerusalem in Israel in 2012, joining already-established research groups at the Smithsonian Migratory Bird Center and the Max Planck Institute for Ornithology, a part of which became the Max Planck Institute of Animal Behavior in 2019.

The new wildlife-tracking tags could not capture the totality of animal movements around the planet as Wikelski hoped ICARUS might: most could affordably transmit data back to scientists only when their wearers stayed within range of cell phone towers, among other limitations. But they did allow scientists to expose how deeply the scale, complexity, and meaning of animal movements had been misunderstood. In every wildlife-tracking project they took on, says Nathan, who directs the Minerva Center, the tags allowed scientists to make discoveries "quite in contrast to the simple explanations we had so far." Giraffes wandered beyond the borders of a national park in Ethiopia, the conservation scientist Julian Fennessy and his team found in GPS tracking studies. Jaguars in the Amazon padded across ranges ten times larger than established by studies conducted with fixed camera traps, the wildlife ecologist Mathias Tobler discovered.

GPS studies challenged conventional understandings of wild animals' roles in seed dispersal and the spread of disease. In a 2009 GPS tracking study, Wikelski discovered that the oilbirds Alexander von Humboldt once condemned as parasites spent so much of their time dropping seeds onto the forest floor that they were "perhaps the most important long-distance seed-disperser in Neotropical forests." Gazelles in Mongolia, a GPS study revealed, could not be responsible for outbreaks of foot-and-mouth disease in livestock: the disease moved five times faster than the gazelles.

Wikelski soon discovered a "physiological ease" in the way animals moved that belied the belabored effort scientists traditionally pictured. In one tracking study, for example, he and his colleagues found that thrushes spent twice as much energy on stopovers as they did while they were in flight. The flying, in other words, was the easy part. In another, his team found "massive" differences in the heart rate of a thrush when migrating compared with when flying in a wind tunnel. The capacity for movement, he says, had been "totally underestimated."

Tracking studies began to endow animal movements with rich new meaning, revealing unexpected links between the movement of disconnected, far-flung species and obscure environmental phenomena. Scientists obtained tantalizing evidence of mysterious animal perceptions, including some that exceeded that of human technology. An unpublished tracking study led by Wikelski in 2011 uncovered correlations between the skittering of goat and sheep up and down the slopes of Mount Etna in Sicily and the intensity of volcanic eruptions, for example, and another tracking study published in 2020 found correlations between the kinetics of farm animals in the Italian village of Capriglia and their distance from the epicenter of earthquakes. In another unpublished tracking study, Wikelski found that the remote desert locations to which storks migrated from thousands of miles away were the same ones where desert locusts emerged, obscure sites that have largely eluded human detection since biblical times. In a study of caribou herds dispersed over thousands of kilometers, the earth scientist Natalie Boelman and her team discovered a correlation that "nobody knew about," Boelman says, between the timing of spring migrations and large-scale ocean-driven climate patterns.

The revolution in wildlife tracking offered a glimpse into the world that ICARUS seeks to reveal. It's one in which geographic borders are porous and migrants make their way across the globe almost effortlessly, like hang gliders on a front. It's one in which movements once deemed episodic are continuous, in which those regarded as rare are common, in which others dismissed as ineffectual are ecologically fundamental. It's a vision of a planet that vibrates with motion.

After nearly two decades, scores of international collaborations, and tens of millions of dollars in funding, Wikelski finally catapulted the ICARUS wildlife-tracking receiver into space. It was built by DLR, the German space agency, and attached to the exterior of the International Space Station by Russian astronauts in 2018. It now orbits the Earth, hundreds of miles above the surface, streaming geographic, environmental, and health data collected from tagged animals across the planet to a ground station in Moscow, and from there to an open-source database called Movebank, which Wikelski and Kays first developed to track ocelots and orchid bees on Barro Colorado Island.

This fall, after refining the manufacture of the tags and the ICARUS software, Wikelski and his colleagues began attaching the tags to wild creatures. Larger tags have been affixed to rhinos, giraffes, zebras, wild dogs, hyenas, and Saiga antelopes; smaller tags to blackbirds. Hundreds of research groups have been lined up to use the tags on their swimming, crawling, and flying subjects—tags whose size Wikelski hopes will drop to just a single gram by 2025, allowing researchers to track small bats and even large insects like dragonflies, butterflies, and desert locusts. As these creatures' faint tangle of tracks thickens and clarifies, the internet of animals blinks to life.

In following the movements of creatures as diverse as dragonflies, koalas, and northern elephant seals, ICARUS may reveal general rules of mobility that are detectable across taxa and habitats and predictable by, say, body size or gait. But some of the most urgent questions ICARUS will answer will revolve around why animals die. Take the yellow-billed cuckoo, for example. The cuckoo's numbers have been shrinking in recent years, but conservation scientists are unsure why. Ornithologists knew they headed to South America in the winter, but just where in the continent remained obscure. A tracking study by Stanley and Marra, as yet unpublished, revealed that the cuckoos congregated in the Gran Chaco, one of the largest and most biodiverse forests in South America. This—as much as or even more than the degraded riparian areas that some scientist blamed—may explain the cuckoos' decline: the Gran Chaco is being rapidly denuded by the expansion of agribusiness. Global wildlife tracking could provide similarly revelatory detail on other declining species, one million of which currently face extinction, according to an assessment by the Intergovernmental Science-Policy Platform on Biodiversity and Ecosystem Services.

Such knowledge will be of immediate practical utility in the urgent task of stalling biodiversity loss.

In the past, scientists acquired such insights by accompanying animals into their wild places, with all the terror and tedium that entails. With ICARUS they will do so by watching blips on a screen and crunching satellite data. But that physical alienation from the living, breathing ferocity of wild creatures, Wikelski says, belies the deeper connection that wildlife tracking allows.

Through the pulses of data streaming from the tags to the ICARUS computers, the wild animals tell us "what they feel, what they see," he says. "It's the closest you can really—not talk to, but at least let the animal talk to you." What we hear could draw them closer to us, before they slip away.

Nature Is Roiled

JEFF GOODELL

Our Summer from Hell

FROM *Rolling Stone*

NOW YOU CAN SEE HOW IT WORKS, this whole climate collapse scenario that writers and scientists have been hollering about
for years. In the space of a few short months, the Pacific Northwest was baked by an extreme heat wave, California was (and still
is) consumed by wildfire and parched by drought, Tennessee was
hit by seventeen inches of rain that caused devastating floods that
killed twenty-two people, a major hurricane flattened towns and
knocked out power for nearly a million people along the Gulf
Coast and then moved north and drowned one of the richest cities
in the world. All of this was in addition to battling a mutating virus
that has already killed more than 645,000 Americans and may or
may not be a preview of a new climate-driven pandemic era.

All this is happening with just 1°C of warming. "This is climate
change," climate scientist Andrew Dessler said in an interview
on CNBC. "It's just a small preview of what is going to happen if
we don't stop emitting greenhouse gases into the atmosphere. We
really need to do that, or we are going to look back on this as the
good old days."

Of course, terrible flooding in Bangladesh and China is also
climate change, as are heat waves in India and Pakistan. But because this wild summer culminated in a flood of the media capital of the world, it has inspired a lot of media coverage and yet
another round of questioning about whether this is the moment
that Americans will begin to truly grapple with the future we have
created for ourselves by our century-long fossil fuel binge. As Chris

Cilizza at CNN put it: "Is this *finally* the moment we wake up to the climate crisis?"

Fifteen years ago, I was out on a research vessel in the North Atlantic and asked a scientist a similar question. We were drilling sediment cores, looking for evidence of past climates in the shells of tiny organisms called forams that lived in ancient oceans. One evening as we sat out on the fantail of the ship, I asked Lloyd Keigwin, a geologist at Woods Hole Oceanographic Institute who was the chief scientist on the trip, what he thought it would take to wake Americans up to the risks of climate change. "When a major hurricane comes along and wipes out a great American city, they will wake up," Keigwin confidently predicted.

Well, we've had Katrina, Sandy, Laura, Ida—I lose track of them all. It's impossible to calculate how many houses have been destroyed, power lines downed, roads washed away, lives upended and lost.

And what has changed? Yes, we have bent down the carbon-emissions curve enough to make the truly apocalyptic climate nightmares (5° C of warming by 2100) less likely. Clean energy prices have fallen precipitously. We elected a president who now talks bluntly and frequently about the climate crisis and has committed to a zero-carbon grid by 2035. Places like Louisiana have invested billions in coastal resiliency projects. Electric bikes and scooters and cars are proliferating. Media-savvy scientists like Michael Mann, Andrea Dutton, Katharine Hayhoe, and Andrew Dessler are speaking ever more clearly about climate risks. Grassroots activist groups are gaining political power and learning how to use it. Climate warriors like former secretary of state John Kerry are jousting in the fields of diplomacy, trying, once again, to convince the Chinese to step up and show leadership in the upcoming international climate negotiations.

This is all good. This is all important. But if this is what it means to "wake up" to the risks of the climate crisis, then we truly are fucked.

I recently moved to Texas, which, according to the 2020 census, is one of the fastest growing states in the nation. I moved to Austin for love, not BBQ or to escape paying state income taxes, and there is much to admire about the state. But climate-wise, the view is bleak. It's highways, strip malls, and big trucks as far as the eye can see. I have visited the big wind farms in the northwest part of the state and there are plenty of venture capital bros driving around in Teslas, but Texas has a governor who is more interested in regulating uteruses and militarizing the border than cutting carbon and preparing for life in a different climate. And just down the road is Houston, the oil capital of the world, where, despite all the talk about economic diversity, fossil fuel still reigns supreme.

And it's not just Texas. The very day New York City was drowning, Senator Joe Manchin wrote an opinion piece for the *Wall Street Journal,* explaining why he would not support spending $3.5 trillion to help slow climate change. This from a politician who reportedly owns millions of dollars in coal stocks and whose state has probably contributed more carbon to the atmosphere than any other. Evidence of cluelessness and short-term greed is everywhere: Last year, billionaire investor Warren Buffett pumped $10 billion into natural gas pipelines. South Florida real estate is still booming, despite the obvious risks of storm surge and sea level rise. What writer and futurist Alex Steffen calls "the brittleness bubble" continues to expand.

Most important, there is zero accountability for the corporations who have made billions of dollars keeping America hooked on fossil fuels while undermining and distorting the urgency of the climate crisis. What price has ExxonMobil and Shell and BP paid for their years of hawking fuels that they knew very well would heat up the planet and cause wreckage and mayhem for generations to come? None, as far as I can tell. There are various lawsuits moving through the courts and activist shareholders are pushing them to think differently about the future, but basically they trashed the planet and got away with it and their only penance is to run commercials about what good citizens they are and what a good job they are doing "innovating for the future." Meanwhile, the *New York Times,* whose climate coverage is exemplary, sees nothing wrong with creating and running ads from Big Oil. As writer Emily Atkin put it: "The NYT stopped shilling for cigarettes. Why won't it stop shilling for fossil fuels?"

The big problem America faces here in the early years of the twenty-first century is that we built our world with the idea that we live on a stable, steady planet. The land is here, the ocean is there, and forever it shall be. The rains will come, but they will be rains like we always knew it to rain. It will get hot, but no hotter than it ever has. For forty years now, we have ignored scientists who were telling us about the risks of dumping CO_2 into the atmosphere and how it could change everything, creating a different planet than humans have ever lived on before.

Now, as the world floods and burns, the price of our willful ignorance and denial is becoming clearer. Are a few devastated towns along the Gulf Coast and waterfalls in the New York City subway system going to be what wakes us up from that? I hope so. But I fear that just as there is no "us," there is also no "waking up." If the pandemic has proved anything, it's that the reservoirs of stupidity and self-destructiveness in the American mind are deeper than even the most cynical among us could have imagined. So maybe the best thing we can do right now is not pretend we will "wake up" to the monstrous reality of our time like some character in a fairy tale. Maintaining a habitable planet is going to be a long hard fight, and if this summer from hell has shown us anything, it's that this fight has only just begun.

KENDRA PIERRE-LOUIS

How Rising Groundwater Caused by Climate Change Could Devastate Coastal Communities

FROM *MIT Technology Review*

FAE SAULENAS DOES NOT want your sympathy.

Saulenas, along with her forty-six-year-old daughter, Lauren, spent last winter—their COVID winter—in Saugus, Massachusetts, in a house without a working furnace. Saulenas is in her seventies. Lauren, because of brain injuries she experienced in the womb, is quadriplegic, blind, and affected by a seizure disorder, among other disabilities. In winter, it's not unusual for overnight temperatures in Saugus to dip into the teens. The two could not long survive without heat, so absent a furnace, they relied on a space heater. But the cost of electricity to power it was $750 in February alone, and it warmed only a single bedroom.

Saulenas doesn't tell this story to engender sympathy but, rather, as a warning. The water table, she says, is rising—seeping into gas lines and corroding furnaces from the inside out. That's what happened to hers. And she wants you to know that if you live anywhere near a coast—even one, two, three miles away—that water might be coming for you, too.

For something you've probably never heard about, rising groundwater presents a real, and potentially catastrophic, threat to our infrastructure. Roadways will be eroded from below; septic systems won't drain; seawalls will keep the ocean out but trap the water seeping up, leading to more flooding. Home foundations will crack; sewers will backflow and potentially leak toxic gases into people's homes.

Saugus is a small town roughly ten miles northeast of Boston. On maps, water is one of its defining features, with the Saugus River and its tributaries meandering through the town and heading through marshland to the Atlantic Ocean. Among those salt marshes, blocked from the Atlantic by the peninsula of Revere Beach, is where Saulenas bought her house in 1975.

Given the proximity to the ocean, the source of her recent woes would seem obvious: sea level rise. Since 1950, sea level in the region has risen by eight inches, and that change has not been linear. The sea is rising faster now than it did a generation ago—about an inch every eight years. But the water that left Saulenas out in the cold did not come from the sea, at least not directly.

Her problems began in 2018, when she lost gas—and thus heat—because of water entering an underground main. It was a problem that would persist, intermittently, for several years. Water would enter the gas main, and her utility, National Grid, would be forced to shut off the gas. National Grid would then try to find where the water was coming from, patch the leak, and pump the water out.

Officially, National Grid has not named the source of the problem. But Saulenas thinks the culprit is groundwater.

Even under normal circumstances, the cast-iron pipes that make up roughly a third of National Grid's infrastructure in Massachusetts are prone to rust and corrosion. She thinks these pipes, which once sat comfortably above the water table, are finding themselves intermittently swamped during seasonal high tides that essentially push up the groundwater. And it's that elevated groundwater that she thinks seeped into the gas main, flooded out her gas meter, and eventually corroded her furnace.

Kristina Hill, an associate professor at the University of California, Berkeley, whom Saulenas reached out to in pursuit of answers, agrees. "She was asking me, is this something that comes from sea-level rise? And obviously, the answer is yes," says Hill.

Hill is one of a number of researchers trying to get the public and policymakers to take the risks of rising groundwater seriously. Unlike rising seas, where the dangers are obvious, groundwater rise has remained under the radar. Hydrologists are aware of the problem and it's all over the scholarly research, but it has yet to surface in a significant way outside of those bubbles. Groundwater rise is only briefly mentioned in the most recent edition of the National Climate Assessment, released in 2018; it's absent from many state and regional climate adaptation plans, and even from flood maps.

A 2021 study in the journal *Cities* found that when coastal cities conduct a climate vulnerability assessment, they rarely factor in groundwater rise. "They talk mostly about sea level rise, storm surges," says Daniel Rozell, an engineer and scientist affiliated with Stony Brook University, who wrote the 2021 paper. "But there haven't been a lot of questions about what's going to happen to the groundwater."

Impacts on existing infrastructure and planned climate adaptations could be catastrophic. Remediation efforts that haven't planned for groundwater rise will be rendered useless. Billions of dollars in infrastructure will need to be upgraded. And it will likely affect an area much larger than what's captured on most flood maps. A 2012 study by researchers at the University of Hawaii that factored groundwater into flood risks found that nationwide, the area threatened was more than twice the area at risk from sea level rise alone.

Any coastal area where "the land is really flat, and the geology is [the kind of] loose material that water moves through really easily," says Hill, is "where this is really going to be a problem." This includes places like Miami, but also Oakland, California, and Brooklyn, New York. Silicon Valley communities like Mountain View are susceptible to groundwater rise, as is Washington, D.C. Worldwide, the area at risk includes portions of northwestern Europe and coastal areas of the United Kingdom, Africa, South America, and Southeast Asia.

"The problem is huge," says Hill. "We've way underestimated the flooding problem."

And because of how groundwater moves, people who are at risk may not know it until it's too late. "One of the most important things about the groundwater is that the rising groundwater level precedes any inundation of the surface," says Rozell. Put another way, we will experience groundwater flooding long before the ocean comes lapping at our front door.

The Water Beneath Our Feet

It might seem puzzling that rising seas could cause groundwater to rise. At first blush the two seem unrelated, but the connection is actually simple. That it has long been ignored reflects our bias toward addressing problems we can easily see.

To understand the link, it first helps to understand a bit about groundwater. The water nestled in sediments underground started as surface water, like rain or snow, and eventually seeped down. A layer of saturated soil rests below a layer of unsaturated soil; the boundary between the two is what's known as the water table. And in many coastal areas this layer of saturated soil, which can be meters thick, rests atop salt water from the ocean. As sea levels rise, the groundwater gets pushed up because salt water is denser than fresh water.

And this isn't the only way that the ocean and groundwater are connected.

"Groundwater normally flows out to the sea," says Rozell. "All along the coast, there's what they call submarine groundwater discharge. You might even notice it if you go to the beach at low tide. If you stand in the water, you might feel really cold water right at the edge, in the sand. And that's groundwater just running out continuously into the ocean."

Thus, any protection designed to keep rising seas from encroaching onto land must also factor in how to let groundwater out.

Arguably the first big study in a prominent scientific journal that looked at what sea level rise might mean to groundwater levels was published in 2012 in the journal *Nature* by researchers Kolja Rotzoll and Chip Fletcher of the University of Hawaii. The study came on the heels of a report by the United States Geological Survey and Yale University researchers who looked at what would happen to groundwater in coastal New Haven, Connecticut, as sea levels rose. In both cases researchers found that the two would rise in concert.

"We looked at well records and found that the water table in the coastal zone goes up and down with the tides," says Fletcher. "And so, we realized there's a direct connection between the ocean and the water table. And as the ocean rises due to climate change, the water table is going to rise and eventually flood the land. So, we're gonna have all these wetlands in urbanized areas and around roads, where we don't really want them. And it turns out this is a form of sea level rise that in many areas is more damaging than what people classically think of as the ocean flowing over the shoreline and flooding."

And we're already seeing the effects.

Danger to Human Health

In talking with experts about groundwater rise, what often comes up is that it's more complicated and harder to adapt to than sea level rise. Any solution to one aspect of the problem can create a cascade of others. Take, for example, something as straightforward as sanitation. Ordinarily, in most parts of the United States, when you flush the toilet one of three things happens, depending on where you live: it goes out to a cesspool, a septic system, or a sewer line. But groundwater rise presents increasing challenges for all three.

Cesspools are essentially concrete cylinders with an open bottom and perforated sides. Especially in coastal areas, the cesspools, which should be dry, instead find themselves constantly inundated, says Josh Stanbro, a senior policy director for Honolulu's city council, who until last January was the city's chief resilience officer. "They're now sort of always wet," he says. Microbes stay alive because they are wet, and because there's so much more water around, they can leach out.

And Honolulu is not the only city with this issue. Miami-Dade County is facing similar problems with septic tanks, which in theory provide a layer of filtration that cesspools do not. But to do that filtration, the systems require a layer of soil two feet deep, and that layer shrinks as water tables rise. Already, 56 percent of the county's systems are periodically compromised during storms. By 2040, estimates suggest, that number will rise to 64 percent. Failed septic systems can contaminate the local aquifers that a community depends on for drinking water.

One work-around is to switch those households and businesses currently on septic or cesspool systems over to sewer lines. In Miami-Dade County, the estimated cost for that shift is $2.3 billion.

Nor are sewer systems a panacea, cautions Berkeley's Kristina Hill. "Most American sewer pipes, both sanitary and storm sewer pipes, are typically cracked, because we do such bad maintenance.

We're like an international joke," she says. "People start conferences in civil engineering in Europe with slides of how bad American systems are, to loosen up the audience." Those cracked sewer pipes let groundwater in. And in places like New York City and Boston, which have what are known as combined sewer systems, water from rain and water from raw sewage mingle, so there's less space in the pipes. This is why as groundwater rises, places like New York City's Jamaica Bay community end up with liquid bubbling up from storm drains during high tide.

Newer cities tend to have systems where rainwater goes into one pipe and sewage into another. But if the pipes are full of groundwater when it rains, there's still nowhere for that rainwater to go. So, in both cases, according to Hill, you'll get more flooding.

There's another way, too, in which rising groundwater can turn our sanitation systems into killers.

"In the Bay Area there's so much legacy contamination under the ground from military use, from the Silicon Valley tech booms—it left a lot of nasty stuff," says Kris May, a coastal engineer and climate scientist who founded Pathways Climate Institute. "And what often happens is we put low-income houses in those areas after they're remediated. But they still leave a certain amount of contamination in the ground, and those regulations were based on no rising groundwater table."

Now the groundwater table is rising. And as it does, it saturates the soil, unlocking contaminants such as benzene. These chemicals are highly volatile, and as gases they can easily find their ways through sewer lines and into homes.

This is the impact of groundwater rise on just one system—sewage. But it could affect many more. Buried electrical lines that aren't properly sealed will short out; foundations will start to heave from the pressure. Some fear that seismic faults could even be put under pressure.

How Water Finds a Way

To protect themselves against rising seas, cities are turning to the same tools they have used for centuries: levees and seawalls. Boston has proposed a 175-mile seawall called the Sea Gates Project. Miami has a proposal for a $6 billion, 20-foot-high seawall. New

York has proposed its own $119 billion, 6-mile-long project called the New York Harbor Storm-Surge Barrier. Homeowners from Florida to California are erecting barriers to keep the ocean out. But the fundamental problem with all these interventions is the same: a seawall holds back the sea, not groundwater.

In some areas, if the underlying ground is relatively impermeable, it is possible to build a seawall or levees that slow groundwater rise. But then you're left with other problems. Recall that water moves toward the ocean. A barrier that stops groundwater from rising with sea level will also keep stormwater from, say, recent rainfall from flowing to the sea.

"If you don't let the water run out to the ocean, then you have to basically pump it over the wall. And that's essentially what the Netherlands has been doing for several centuries," says Stony Brook's Rozell. But this, too, can create problems, because so many of the places these seawalls are working so hard to save—much of Lower Manhattan, large parts of San Francisco and Boston—were built on wetlands, landfill, or both. "If they pump, the land is going to sink," says Hill.

And even if cities were willing to pursue such a path, not every place can. "There are lots of conditions where you can pump all day long and the water table won't go down," says the University of Hawaii's Fletcher.

Recall that groundwater is water that makes its way into the spaces, or pores, in sediment. In some places, like Miami, "the pores are so large that you're just pulling in water from the estuary from the ocean," says Fletcher. "You can pump as hard as you want, and it just keeps coming in from an endless body of water"— the sea.

Planners are often oblivious to the problem. In 2009, the Maldives, a low-lying island nation, held the world's first underwater cabinet meeting to draw attention to the harm big climate polluters, like the United States, were perpetuating through climate inaction. The message was clear: You're drowning us. These days, already dealing with the consequences of rising seas, the country is consolidating its outer island communities onto a new island called Hulhumalé. It's designed to withstand sea level rise. But the project did not factor in rising water tables.

"They did not understand that the water table will rise with sea level rise," says Fletcher. If the sea rises only two more feet—

which some estimates say will happen as soon as 2040—most of this brand-new island will be uninhabitable wetland.

When he explained this to the project's lead designer, "he just stared at me—he was speechless. It's like he couldn't comprehend what I was saying," Fletcher says. "All the billions of dollars they had spent on this thing, and they didn't build it high enough."

Eroding Away History

There is at least one place where you can see people reckoning with rising groundwater in close to real time. Strawbery Banke Museum is in Portsmouth, New Hampshire, near the banks of the Piscataqua River, just a few miles from the Atlantic Ocean. The buildings were preserved to let us see three centuries into the past, but they are also giving us a glimpse into the future. Some of the structures, including the city's second-oldest house, are flooding from below.

"We're getting these super tides, king tides, that elevate the water over two feet higher than typical. And so, we're starting to see this water get into our basements," said Rodney D. Rowland, Strawbery Banke's director of facilities and environmental sustainability, on a tour of the museum in late September. When you crouch down in basements with their ceilings too low for most adults to stand, it's easy to see the watermarks from past groundwater incursions.

The museum has taken a two-pronged approach. The first element is educating the public. "One of the exciting things that we're gonna add is a kiosk that is attached to sensors that were placed in the ground around the museum," said Rowland. "And they will track the movement of the groundwater, [plus] salinity, temperature, water height. And so, visitors will see that there's water under their feet."

But the museum also needs to preserve the buildings. And that goal must now be balanced with the fight against rising water. In one of the houses, "we made the decision to take out what was called a summer kitchen," said Rowland. "There was a hearth down there where they cooked in the summertime. We took it out, and we put in a granite block." They had to do that because the old hearth was acting like a candlewick, drawing water from the basement into the rest of the structure.

"So now the rest of the chimneys are preserved," he added. "The water can't get through that. But we lost that piece of history. And this is going to be a constant battle with how much are we going to lose to save what we can."

In some ways Rowland is lucky. His state, New Hampshire, is at least aware of the risk of groundwater rise and is factoring it into plans. But New Hampshire is an exception. Many other states, with more extensive coastlines, are going to have to face the issue in the coming years as not only buildings but lives are threatened by this unseen risk.

Less than fifty miles down the coast in Saugus, Fae Saulenas plans on leaving for higher ground—but not without making some noise. She's written legislators, National Grid, and the press to try to draw attention to the issue. "Groundwater is really important to me. And it's important to me not only because it has affected my life profoundly, but because I think it has the capacity to affect millions of people," she says. "And nobody's prepared, and nobody's paying attention."

MARK ARAX

How We Drained California Dry

FROM *MIT Technology Review*

THE WIND FINALLY BLEW the other way last night and kicked out the smoke from the burning Sierra. Down here in the flatland of California, we used to regard the granite mountain as a place apart, our getaway. But the distance is no more. With all those dead pine trees in thrall to wildfire, the Sierra, transmuted into ash, is right outside our door.

We have learned to watch the sky with an uncanny eye. We measure its peril. Some days, we breathe the worst air in the world. On those few days when we can walk outside without risking harm to our lungs and brains, we greet one another with new benedictions. May the shift in winds prevail, I tell my neighbor. May there be only the dust clouds from the almond harvest to contend with. In the meantime, I don't dare quiet the turbo on my HEPA filters, hum of this new life.

The most brutal of summers in the San Joaquin Valley has come to a rest at last. Since June, the temperature has broken the 100°F mark for sixty-seven days, a new record. Drought won't let go its grip on the land. Eight of the past ten years have been ugly dry. This October morning, after a month holed up, I decided to leave my house in the suburbs and roam the middle of California, the irrigated desert at its most supreme. Out in the country, I smell fall in the air. To celebrate its arrival, I'm going to visit an old friend, a farmer named Masumoto, who has eighty acres in Del Rey and is putting the last of his raisins in a box.

There is no way to make this drive out of Fresno at harvest's end, through the dog-tired fields of the most industrialized farm belt in the world, without thinking about water: the idea of it, the feel of it; the form as it falls from the sky as rain and snow, that man captures with his invention and implementation, his magic and plunder, the dam, the ditch, the canal, the aqueduct, the pump, the drip line; the water that gives rise to every animate and inanimate thing that now stretches before my eyes, the vineyard, orchard, cotton field, and housing tract; the water whose too much can destroy us, whose too little can destroy us, whose perfect measure of our needs becomes our superstition and story.

You should know that I have written about the matter of California and water a few times before, and I'm not above borrowing from old refrains. In my hunt for new words, I have driven Highway 99 a thousand times through a valley that geologists call the most altered landscape by human hands in history. I now see the gashes of fresh alterations. What has been done here, by any means necessary, has been done for the want of water.

The taking of California was no small project. It relied on the erasure of the most prolific flowering of Indigenous people in the United States. The civilization standing in the way was at least ten thousand years in the making and three hundred thousand strong. They were Yokuts, Maidu, Miwok, Klamath, Pomo, Chumash, and Kumeyaay, to name a few. Looking back at the fevered pace of our footprints over the past 175 years, we tend to idealize the modesty of theirs. And yet it is more than likely true given their numbers, given the bounty and heft of the land, that they did not war with one another over its prize. They lived light on the Earth. They moved when nature moved. Flood took them to one place, drought another. When the forest load needed thinning, the fires they set burned brush and lower branch and quickly smothered out.

As genocides go, the wiping clean of California's Indigenous culture was protracted, playing out in three acts: Spanish mission, Mexican occupation, American settlement. The atrocities were only as efficient as the tools of the time—blanket, smallpox, syphilis, torch, knife, Colt .45—allowed. First came the robed Franciscans led by Father Serra, slaver and saint, whose possession of the Indian body gave him the workforce to erect the first crude dams and canals that took rivers to places they'd never been: his twenty-one missions, from San Diego to Sonoma. At the Mission San Gabriel, the catch of water grew a profusion of grains, vegetables, exotic fruits, and the 170-acre Las Vina, "mother vineyard."

Next came the dons from Mexico, freed from Spain's yoke, whose dalliance with California lasted but a quarter century, from 1821 to 1848. Blending European, Mexican, and American lineages, they called themselves Californios. Rather than tame California's many states of nature, they amassed millions of acres and tamed themselves. On far-flung rancherias, they slaughtered a calf a day to feast on, drank vast quantities of wine and brandy, and threw royal weddings in which daughters who'd been locked away in finishing schools all their lives finally came out into the sun. In a moment of goodwill, they pledged that the mission lands, and their flow of water, would be turned over to the remaining Natives, but the pledge never amounted to a thing.

American settlers had been nosing around for decades—mountain men, fur trappers, scouts, and surveyors. When they finally made their intentions known, in the summer of 1846, the government standing behind them grabbed the western edge of a continent, one thousand miles long, without firing an official shot. What are a people to do when the land they conquer covers eleven regions of topography and ten degrees of latitude, where rain measures 140 inches on one end and two inches on the other end? Another people might have taken the stance that each region ought to exist within its own plenitude and limit. These people drew a line around the whole, declared it one state, and began their infinite tinkering to even out the difference.

Manifest destiny would have had its way with California, sure and steady, but the shout of gold, in 1848, was heard around the world. Gold's cataclysm was a force of a different magnitude. Overnight they sailed ashore by the tens of thousands, mad miners from all over the globe, most of whom had never mined a day in their lives. They went at mountain and river with claws. Mining gold, they discovered, was mining water on an industrial scale.

"Water! Water! Water!" shouted James Mason Hutchings, an Englishman who published a quarterly of unparalleled excellence, *Hutchings' Illustrated California Magazine,* in the 1850s. "Not water to drink, for that can be found bubbling up on every mountaintop, but water to work with. Working men dig gold. Gold, thus dug, would be put in circulation. That circulation would give prosperity. We will therefore, with the same language as the horseleech, cry, 'Give, Give,' but let the gift be Water! Water! Water!"

By the time the great deluge of 1862 rained down, Hutchings's magazine was no more. It would be left to William Brewer, who studied at Yale and came west to survey California's natural resources, to describe what the floodwaters had done. "Nearly every house and farm over this immense region is gone," he wrote. "America has never before seen such desolation by a flood as this has been." Brewer had come to recognize the Californian's peculiar fortitude to outlast everything: "No people can so stand calamity as this people. They are used to it."

The people forgot about flood with the same nonchalance that they forgot about drought. Their failure of memory became a strange resilience. They went back to their digging with newfound lust. They erected six thousand miles of ditches and built a dam one hundred feet tall. The flows of Northern California rivers were now dictated by a handful of industrialists. To reach the deeper veins of gold, they invented hydraulic cannons that shot out water at such force that it blew the walls off mountains. Into the rivers washed the tailings, more than a billion cubic yards of boulder, rock, pebble, and mud. Tens of thousands of acres of new crops planted in the alluvial plain began to choke on the retch of the mines.

As to the future of California, the industrialists who lived atop San Francisco's Nob Hill had a choice to make: gold or grain? Isaac Friedlander, six foot seven and three hundred pounds, whose stride was said to be that of two men, who had made his fortune by cornering the market on flour for the mining camps, snatched one million acres of valley soil for practically nothing. He became the Wheat King.

I am sailing across the desert, that is true, but it isn't the Mojave. The San Joaquin Valley, 260 miles long and 50 miles wide, qualifies as desert only by measure of average rain—less than 10 inches a year. Five rivers, two of them mighty, run down from the Sierra across its breadth. The best of the dirt, a loam that blends sand and clay, grows beets the size of an ogre's head. The sun shines 280 days a year, and the sky doesn't generate any rain from May to September. The blanket of fog that sets down in winter holds the chill of hibernation close to the fruit and nut trees. The importance of these chilling hours was a lesson my father's father, Aram Arax, a poet-farmer, thought I should know: "The apricot is a picky thing. It has to feel the kiss of the death in winter to hold on to its fruit in spring." He would need to go back to the Mediterranean, he told me, to find a clime where all manner of vegetables, fruits, nuts, and grains grew with such ease.

The 49ers who had made their way down the hill knew what to do with this fecundity. So did the cotton growers from the South who were chased off their plantations by the boll weevil. They corralled the rivers with a lattice of ditches and made them run backward. They drained dry the great inland marsh and Tulare Lake, too, the largest freshwater body west of the Mississippi. They wiped out the last of the elk, antelope, and mustang and emptied the sky of geese. They flattened the hillock and hog wallow with the Fresno Scraper and turned 6 million acres into tabletop. That's how the water of furrow irrigation glided.

Their seize of the snowmelt—"first in line, first in right"—had no parallel in agriculture. They did not take half the flow of the rivers. They did not take three-quarters. By the time the farmers were done, they had taken nine out of every ten drops. When their garden was ready for showing, their promotional brochures fairly boasted, "Fresno County: A Wonderfully Prosperous District in California. The Land of Sunshine. Fruits and Flowers. No Ice. No Snow. No Blizzards. No Cyclones."

It would be easy to dismiss the lure of such hype. But word of their feat—"the first great experiment in irrigation by the Anglo-Saxon race"—reached all the way to Istanbul, to the attic where my grandfather Arax was hiding from the Turks in 1918. His uncle, who had lost his wife and children in the massacres and had fled to Fresno, was writing him letters describing an Eden in a valley at the edge of the Sierra: "You must see it with your own eyes to believe it."

My grandfather was plotting his way to the Sorbonne, to study French literature and become a writer, but the letters kept coming, each one more full of sadness and hope than the one before. In the summer of 1920, after a seven-thousand-mile-long journey, he found himself at the train depot in downtown Fresno. Nephew and uncle, survivors of genocide, hugged and climbed into a gleaming Model T Ford and rode from river to river, across an expanse already known as the "Raisin Capital of the World." They passed grapes, peaches, and plums and lingered on twelve thousand acres of figs that a Kansas preacher was planting in the red hardpan. My grandfather, awestruck, kept muttering the same words: "Just like the old land."

As I approach the Kings River—emptied of river, nothing but sand—I can hear the words he used to describe our last farm, the one *embroidered* with pomegranate trees that my father, Ara, and his brother, Navo, to my grandfather's regret, sold a few years before I was born. I grew up in the suburbs not a dozen miles removed from those sixty acres, but it might as well have been an ocean away, for who we were and what we had done to make the desert bloom wasn't a topic we discussed.

We had the Cotton King, Grape King, Melon King, and Tomato King right in our midst, men who possessed the lion's share of our water, but how this dominion had happened remained a civic mystery. Irrigation canals full of snowmelt knifed through our neighborhoods, but it never occurred to me to ask where the water was coming from, to whom it was going, and by what right. The canals were completely unfenced, and one or more children of the Mexican farmworkers, looking to cool off in summer, drowned in them every year. "Don't go next to those canals," my grandmother Alma warned. "If you fall in, they won't fish you out. They won't stop the flow until the harvest is over."

The new land was nothing like the old land.

Not a year after my grandfather arrived, the raisin went bust. The Armenian and Japanese farmers had planted so many grapes to dry into raisins that Sun-Maid couldn't sell half of them. Who would buy the other half became a question of such wonderful theater, tragic and comic, that even Fresno's sage, William Saroyan, would weigh in. If we could only persuade every mother in China to put a single raisin in her pot of rice, we'd have the glut solved, he mused.

Just as the bust hit, the great drought of the 1920s hit, too, revealing the folly and greed of California agriculture. It wasn't enough that the farmers had taken the five rivers. They were now using turbine pumps to seize the aquifer, the ancient lake beneath the valley. In a land of glut, they were planting hundreds of thousands more acres of crops. This bigger footprint wasn't prime farmland but poor, salty dirt beyond rivers' reach. As the drought worsened, the new farms were extracting so much water out of the ground that their pumps couldn't reach any lower. Their crops were withering.

A cry went out from the agrarians to the politicians: "Steal us a river." They were eyeing the flood flows of the Sacramento River up north. If the plan sounded audacious, well, just such a theft had already been accomplished by the City of Los Angeles, reaching up and over the mountain to steal the Owens River.

This is how the federal government, in the 1940s, came to build the Central Valley Project, damming the rivers and installing mammoth pumps in the Sacramento–San Joaquin Delta to move water to the dying farms in the middle. This is how the state of California, in the 1960s, built the State Water Project, installing more pumps in the delta and a 444-mile-long aqueduct to move more water to grow more farms in the middle and more houses and swimming pools in Southern California.

This is how we've come to the point today, during the driest decade in state history, that valley farmers haven't diminished their footprint to meet water's scarcity but have added a half million more acres of permanent crops—more almonds, pistachios, mandarins. They've lowered their pumps by hundreds of feet to chase the dwindling aquifer even as it dwindles further, sucking so many millions of acre-feet of water out of the earth that the land is sinking. This subsidence is collapsing the canals and ditches, reducing the flow of the very aqueduct that we built to create the flow itself.

How might a Native account for such madness?

No civilization had ever built a grander system to transport water. It sprawled farmland. It sprawled suburbia. It made rise three world-class cities, and an economy that would rank as the fifth largest in the world. But it did not change the essential nature of California. Drought is California. Flood is California. One year our rivers and streams produce 30 million acre-feet of water. The next year, they produce 200 million acre-feet. The average year, 72.5 million acre-feet, is a lie we tell ourselves.

I am sitting on the porch of a century-old farmhouse, eating kebabs and pilaf with David "Mas" Masumoto. We're looking out in near silence at his eighty acres of orchards and vineyards not far from the Kings River. His small work crew has gone home. His wife, Marcy, is doing volunteer work overseas, and their three dogs, all stinking, know no bounds. The whole place looks exhausted, like a farm where the farmer has died. But Mas, nearing sixty-eight, is as alive as ever.

We got to know each other twenty-five years ago on the occasion of his first book, *Epitaph for a Peach,* a memoir about a farm passed down from father to son and the son's determination not to plow under an old variety of the fruit. The heirloom was called

Sun Crest, and it had fallen out of favor with the market because it bruised too easily. Golden, sweet, and juicy, it was worth saving, Mas thought. "You take one bite, and it throws you back in time," he had told me then. "Fruit is memory."

I hadn't heard a farmer talk that way since my grandfather, and so I wrote a story about him in the *Los Angeles Times,* and he handed me a young Sun Crest to plant in my own backyard, and it bore so many peaches next to the swimming pool that my wife, after our divorce, declared the tree a "mess" and pulled it out. Mas, on the other hand, had saved the peach. Chef Alice Waters, for one, read his book and started serving Sun Crests, all by themselves, as dessert at Chez Panisse.

He points to a spot in the orchard where they're still standing, more gnarled and weather-beaten but still producing. He counts himself among the lucky. His father, Takashi, chose this land well. It sits inside an irrigation district with an early call on the river. Even in low-runoff years, his water table gets recharged.

"We're irrigating right now, matter of fact," he says. "The water table has dropped some, but out here that means we're sitting at seventy feet [deep]. Up and down the valley, it doesn't get much better than that."

"How'd the harvest go?

"It's the middle of October, and it's still going," he says with disbelief.

Talking about the weather with a farmer isn't like talking about the weather with anyone else. It's prying into the soul of things. I venture the opinion that this long dry spell isn't only California returning to drought form. It's climate change hitching onto drought, creating an altogether new havoc. Mas isn't like most farmers. He grows his fruit organically and drives a Prius. "Climate" and "change" are words he speaks together.

"I've seen things this harvest I've never seen before," he says.

We finish our kebabs and walk the century-old rows. The Thompson seedless vines look ready to kiss winter and fall asleep. But the amber grapes laid out on paper trays in the terraced loam are only half baked. I know the rhythm, and the rhythm is off. Thompsons are put down in early September to avoid fall's first rain. It takes but twelve days for the valley sun to wrinkle a grape into a raisin. Mas's raisins are a month late in drying. They've already been rained on once.

"It's a mystery," he says.

He bends down into the crouch that raisin farmers assume when they are about to examine their crop. He sifts through the bunches with his sunburned hands, feeling for that sticky. He puts a couple in his mouth, feeling for that chewy. It's not there.

"Not a raisin yet? How do you figure?" I ask.

He looks to the sky. "This summer was record hot. They should have ripened right up. But the sun didn't shine the same."

I don't know what he means.

"All that smoke and ash from the forest fires. It changed the rays, I figure. It bent them somehow. They didn't come through the same."

I nod and keep listening. He is talking about nature's cycle. Drought helped kill the trees in the forest. Desiccated by thirst, they were whittled out by bark beetles. Lightning lit that kindling. Kindling became smoke and ash. Smoke and ash occluded the sky. This slowed the ripening of grapes on the vine. This slowed the baking of grapes into raisins.

Thanks to the wind, the sky is now clear, but it's too late. October has changed the angle of the sun hitting the rows.

"We've lost our oven," he says. "I'll likely be sending these raisins to the mechanical dryer. That's never happened before. They won't taste the same."

It was hard to find a sweeter spot on Earth for farming. Mas had the soil, he had the river, he had the aquifer, and he had the sunshine, or at least he thought he did. He did not have the science to explain it, but climate change had found him, too.

"I think of our farm as being alive," he says. "Nature is alive. Climate is alive. Is the idea to try and kill it? I'm not saying we don't fight climate change as a society. We have no choice but to. But out here, it's folly trying to control nature."

We walk past the giant concrete standpipe, filling up with water that will give a last drink to the farm before winter. He talks proudly about his daughter, Nikiko, and his son, Korio, who will take over these acres sooner rather than later.

"Out here, everything is going to take time," he says.

We hug goodbye. I get into my little Chevy, turn on the electrical engine, and drive home through the dust. The pomegranates are turning red, and I can't help thinking: How much time do we have?

LISA SONG AND JAMES TEMPLE

The Climate Solution Actually Adding Millions of Tons of CO2 into the Atmosphere

FROM ProPublica/*MIT Technology Review*

ALONG THE COAST of Northern California near the Oregon border, the cool, moist air off the Pacific sustains a strip of temperate rainforests. Soaring redwoods and Douglas firs dominate these thick, wet woodlands, creating a canopy hundreds of feet high.

But if you travel inland the mix of trees gradually shifts.

Beyond the crest of the Klamath Mountains, you descend into an evergreen medley of sugar pines, incense cedars, and still more Douglas firs. As you continue into the Cascade Range, you pass through sparser forests dominated by Ponderosa pines. These tall, slender trees with prickly cones thrive in the hotter, drier conditions on the eastern side of the state.

All trees consume carbon dioxide, releasing the oxygen and storing the carbon in their trunks, branches, and roots. Every ton of carbon sequestered in a living tree is a ton that isn't contributing to climate change. And that thick coastal forest can easily store twice as much carbon per acre as the trees deeper inland.

This math is crucial to determining the success of California's forest offset program, which seeks to reduce carbon emissions by preserving trees. The state established the program a decade ago as part of its efforts to combat climate change.

But ecology is messy. The boundaries between forest types are nebulous, and the actual amount of carbon on any given acre depends on local climate conditions, conservation efforts, logging history, and more.

California's top climate regulator, the Air Resources Board, glossed over much of this complexity in implementing the state's program. The agency established fixed boundaries around giant regions, boiling down the carbon stored in a wide mix of tree species into simplified, regional averages.

That decision has generated tens of millions of carbon credits with dubious climate value, according to a new analysis by Carbon-Plan, a San Francisco nonprofit that analyzes the scientific integrity of carbon removal efforts.

The offset program allows forest owners across the country to earn credits for taking care of their land in ways that store or absorb more carbon, such as reducing logging or thinning out smaller trees and brush to allow for increased overall growth. Each credit represents one metric ton of CO_2. Landowners can sell the credits to major polluters in California, typically oil companies and other businesses that want to emit more carbon than otherwise allowed under state law. Each extra ton of carbon emitted by industry is balanced out by an extra ton stored in the forest, allowing net emissions to stay within a cap set by the state.

As of last fall, the program had produced some six dozen projects that had generated more than 130 million credits, worth $1.8 billion at recent prices.

While calculating the exact amount of carbon saved by preserving forests is complicated, California's logic for awarding credits is relatively straightforward.

The Air Resources Board establishes the average amount of carbon per acre stored in a few forest types spanning large regions of the United States. If you own land that contains more carbon than the regional average, based on a survey of trees on your site, you can get credits for the difference. For example, if your land holds the equivalent of one hundred tons of CO_2 per acre, and the regional average is 40 tons, you can earn credits for saving 60 tons per acre. (This story will refer to each ton of CO_2-equivalent as a ton of "carbon.") You must also commit to maintaining your forest's high carbon storage for the next hundred years.

These regional averages are meant to represent carbon levels in typical private forests. But the averages are determined from such large areas and such diverse forest types that they can differ dramatically from the carbon stored on lands selected for projects.

Project forests that significantly exceed these averages are frequently earning far more credits than the actual carbon benefits they deliver, CarbonPlan found.

This design also incentivizes the developers who initiate and lead these projects to specifically look for forest tracts where carbon levels stand out above these averages—either due to the site's location within a region, its combination of tree species, or both.

CarbonPlan estimates the state's program has generated between 20 million and 39 million credits that don't achieve real climate benefits. They are, in effect, ghost credits that didn't preserve additional carbon in forests but did allow polluters to emit far more CO2, equal to the annual emissions of 8.5 million cars at the high end.

Those ghost credits represent nearly one in three credits issued through California's primary forest offset program, highlighting systemic flaws in the rules and suggesting widespread gaming of the market.

"Our work shows that California's forest offsets program increases greenhouse gas emissions, despite being a large part of the state's strategy for reducing climate pollution," said Danny Cullenward, the policy director at CarbonPlan. "The program creates the false appearance of progress when in fact it makes the climate problem worse."

The Air Resources Board defended the program and disputed the central thesis of the study.

"We disagree with your statement that landowners or project developers are gaming the system or that there are inflated estimates" of greenhouse gas reductions, Dave Clegern, a spokesperson for the Air Resources Board, said in an email. Each version of the offset rules "went through our robust public regulatory review process," with input from the forestry industry, academia, government agencies, and nonprofits, he added.

California's forest offset program is the largest in the country that is government-regulated. Other forest offset programs are voluntary, allowing businesses or individuals to purchase credits to shrink their environmental footprint.

CarbonPlan's study comes days after the Washington state legislature moved a cap-and-trade bill with an offset program to the governor's desk for approval. Oregon has also debated in recent months establishing a carbon market program that would emulate California's policy. In Washington, D.C., the Biden administration has signaled growing interest in harnessing forests and soil to draw down CO_2. Businesses, too, increasingly plan to rely heavily on trees to offset their emissions in lieu of the harder task of cutting corporate pollution.

Forest offsets have been criticized for a variety of problems, including the risks that the carbon reductions will be short-lived, that carbon savings will be wiped out by increased logging elsewhere, and that the projects are preserving forests never in jeopardy of being chopped down, producing credits that don't reflect real-world changes in carbon levels.

But CarbonPlan's analysis highlights a different issue, one interlinked with these other problems. Even if everything else about a project were perfect, developers would still be able to undermine the program by exploiting regional averages.

Every time a polluter uses a credit that didn't actually save a ton of carbon, the total amount of emissions goes up.

Far from addressing climate change, California's forest offsets appear to be adding tens of millions of tons of CO_2 into the atmosphere on balance, undermining progress on the state's long-term emissions goals.

"When you strip away all the jargon, you're left with a faulty set of assumptions that leave the door wide open to issuing meaningless offset credits," said Grayson Badgley, a postdoctoral fellow at Black Rock Forest and Columbia University, and the lead researcher on the study.

Cherry-Picking

CarbonPlan provided ProPublica and *MIT Technology Review* full and exclusive access to their analysis as it was being finalized. As part of that process, the news organizations sent the report to independent experts for review. The organizations also interviewed landowners, industry players, and scientists and reviewed hundreds of pages of documents, including the project plans submitted by developers. CarbonPlan collaborated on the study with academic experts from the University of California–Berkeley, Columbia University, and other institutions.

The study itself wasn't designed to assess whether developers or landowners are intentionally cherry-picking sites that stand out from regional averages, stating only that the system "allows for" developers to select such land. But the researchers themselves say that the level of excess crediting and the clustering of projects in certain areas suggest that industry players have gamed the system.

One form of cherry-picking identified by the researchers involves geographic boundaries. In the case of Northern California, the state's offset program established a dividing line that separates that coastal strip of redwoods and Douglas firs from an inland region that spans more than 28,000 square miles.

The board's rules state that tall mixed-conifer forests in the coastal region store an average of 205 tons of carbon per acre. For the neighboring inland region, the agency set the corresponding regional average at 122 tons per acre. The figure is lower because it includes more trees with less carbon, such as Ponderosa pines, which dominate the eastern end of the inland region and are all but absent on the coast.

But where the two regions meet, the forest on either side is virtually identical in many places, storing similar amounts of carbon. That means a project developer can earn far more money by choosing a site just east of the border, simply because they can compare the carbon in their forest against a lower regional average. For instance, maintaining a 10,000-acre forest of coastal redwoods and Douglas firs with carbon levels of 200 tons per acre could earn zero credits west of the line, or 624,000 credits east of it. The choice is between no money and more than $8 million.

To claim the most credits possible, for the full difference between the carbon on their land and the regional averages, developers or landowners must show that it's legally and financially feasible to log down to those regional averages. The averages are effectively a stand-in for the way that similar forests are typically managed in an area.

A dozen projects are located in Northern California, almost entirely lined up along the western edge of the inland zone where the carbon-rich trees are juxtaposed against the lower regional average.

"What we're seeing is developers are taking advantage of the fact that the big stuff and the scrubby stuff have been averaged together," Badgley said.

Once an offset project developer and landowner decide to work together, the developer will generally shepherd them through the process in exchange for a fee or share of the sales of the credits generated—an arrangement that can be worth millions of dollars.

One of the most prolific project developers in the California system is an Australia-based timberlands investment company called New Forests. The company and its affiliates have worked on eight projects located almost entirely along the advantageous side of the border, as well as six elsewhere. CarbonPlan, in a separate analysis done for news organizations that wasn't included in the study, found that nearly all earned dubious credits, adding up to as much as $176 million worth.

A large share of those credits came from a single project outside California that profited from a glaring mistake in the rules. New Forests' affiliate, Forest Carbon Partners, helped the Mescalero Apache Tribe develop a forest offset project in New Mexico. The project earned 3.7 million credits worth more than $50 million, largely because it was located in an area where the Air Resources Board had set an erroneously low regional average.

Another form of cherry-picking involves tree species: developers can seek out tracts with particular trees that store far more carbon than the surrounding region.

According to the study, one project in Alaska consists almost entirely of giant Sitka spruces, yet the local regional average was calculated from a wide mix of trees, including species like cottonwoods that store far less carbon. The project earned significantly more credits than it should have due to the flaws in the system, the study said. The project owner didn't return requests for comment.

Preserving especially carbon-rich forests is good for the climate, in and of itself. But when the trees in the project area bear little resemblance to the types of trees that went into calculating the regional average, it exaggerates the number of credits at stake, CarbonPlan's study found.

Mark Trexler, a former offsets developer who worked in earlier U.S. and European carbon markets, said the board should have anticipated the perverse incentives created by its program.

"When people write offset rules, they always ignore the fact that there are a thousand smart people next door that will try to game them," he said. Since the board set up a system that "incentivizes people to find the areas that are high-density, or high-carbon, that's what they're going to do."

To estimate the extent of over-crediting in California's program, CarbonPlan calculated its own version of regional averages for each project. The researchers drew on the same raw data used by the Air Resources Board, but only used data from tree species that more closely resemble the particular mix of trees in each project area.

In total, seventy-four such projects had been established as of September 2020, when CarbonPlan began its research. CarbonPlan was able to study sixty-five projects that had enough documentation to make analysis possible. All received credits for holding more carbon than the regional average.

The researchers found that the vast majority of projects were over-credited, but about a dozen would have received more credits under CarbonPlan's formula. Those included two New Forests projects, which would have earned as much as an additional 165,000 credits.

The news organizations sent officials at the Air Resources Board a copy of the study and its detailed methodology weeks before publication. Clegern declined multiple requests to interview board staff and responded only in writing.

He did not address CarbonPlan's calculations. "We were not given sufficient time to fully analyze an unpublished study and are not commenting further on the authors' alternative methodology," he wrote.

The outside scientists who reviewed the research on behalf of ProPublica and *MIT Technology Review* praised the study.

"It's a really analytically robust paper and it answers a really important policy question," said Daniel Sanchez, who runs the Carbon Removal Laboratory at UC-Berkeley. While close observers are well aware of numerous problems with California's forest offset rules, "they're revealing a deeper set of serious methodological flaws," he said.

None of the reviewers pointed out any major technical or conceptual flaws with the paper, which has been submitted to a journal for peer review.

*

"A Significant New Commodity Market"

In early 2015, an offsets nonprofit hosted a webinar highlighting how Native American tribes could participate in California's program.

One speaker was Brian Shillinglaw, a Stanford-trained lawyer and managing director at New Forests who oversees the company's U.S. forestry programs. The company manages the sale of carbon credits, sells timber, and on behalf of investors manages more than 2 million acres of forests globally, a portfolio it values at more than $4 billion.

New Forests also manages its affiliate, Forest Carbon Partners, on behalf of an institutional investment client it declined to name. Forest Carbon Partners finances offset projects and shepherds landowners through the process of applying for California's offset program.

"The bottom line is the California carbon market has really created a significant new commodity market," Shillinglaw said during his presentation. He said the program is something "many Native American tribes are very well situated to benefit from, in part due to past conservative stewardship of their forests, which can lead to significant credit yield in the near term."

Translation: because many tribes have logged less aggressively than their neighbors, their carbon-rich forests were primed for big payouts of credits. Under Shillinglaw, New Forests and Forest Carbon Partners have helped to secure tens of millions of dollars' worth of credits for Native tribes.

Among the thirteen New Forests projects that CarbonPlan researchers were able to analyze, between 33 percent and 71 percent of the credits don't represent real carbon reductions. That's nearly thirteen million credits at the high end.

"Although we cannot prove that New Forests acted deliberately on the basis of our statistical analysis, in our judgment there is no reasonable explanation for these outcomes other than that New Forests knowingly engaged in cherry-picking behavior to take advantage of ecological shortcomings in the forest offset protocol," said Badgley, the lead researcher.

New Forests managed the first official project in California's program, registering 7,660 acres of forest land on or near the Yurok Reservation, which runs more than forty miles along the

Klamath River near the top of that West Coast cluster of projects. The state issued more than seven hundred thousand credits to the project for its first year, worth $9.6 million at recent rates.

State officials have pointed to the tribe's participation as a triumph of the program. In 2014, the board released a promotional video that showed the meticulous work of measuring trees in the Yurok project. James Erler, the tribe's then-forestry director, explained how offsets enabled the tribe to reduce logging. Near the end of the video, Shillinglaw appears in a sunlit forest, wearing a collared shirt and a New Forests–branded jacket.

"It's a beautiful watershed," Shillinglaw said over footage of a running stream and an elk standing before a thicket of trees. "This is the Yurok Tribe's ancestral homeland, and in part due to the carbon market will be managed through a conservation approach."

CarbonPlan estimates the project earned more than half a million ghost credits worth nearly $6.5 million.

Here's why the researchers say it was over-credited:

The boundary dividing California's coastal and inland regions runs through the middle of the reservation. The carbon-rich forests on either side of that line are similar, filled with large Douglas firs like most of the coastal region. But more than 99 percent of the forest designated for preservation falls within the inland zone, where average carbon levels are much lower. The fact that the project was located in the most carbon-rich area of that zone enabled the landowners to earn an exaggerated number of credits.

At least one person involved in the Yurok Tribe's forest offset efforts was aware of how geographical choices swing the credits that can be earned.

Erler said during a 2015 presentation at a National Indian Timber Symposium that the tribe had the "distinct pleasure" of having the boundary run through its territory.

"You can take the same inventory data and apply it to the California Coast"—the region to the west—"and it doesn't come out with the same numbers as you do if you cross the street," Erler said at the conference, captured in a YouTube video posted to the Intertribal Timber Council's channel. "Vegetation may be the same, but it changes."

Badgley said that while the researchers can't speak to the intentions of any actors involved, it's clear that this project "benefited from over-crediting and that the Yurok Tribe's forester was aware how the specific aspects of the protocol rules our study criticizes led to beneficial outcomes."

Erler didn't respond to a list of emailed questions.

In an emailed statement, Yurok spokesperson Matt Mais said that the property was the only land the tribe had available to enroll at the time and strongly denied the tribe engaged in any sort of gaming of the system. He didn't respond before press time to a subsequent inquiry asking why the rest of the tribe's land wasn't available for the offset program.

Over the last decade or so, the tribe has slowly reacquired tens of thousands of acres of its ancestral territory, in and around the watershed of Blue Creek and other streams that sustain migrating salmon, from the Green Diamond Resource Company, a major Seattle-based timber business. The complex, multistep land deals were done in partnership with the nonprofit Western Rivers Conservancy and financed through government grants, philanthropic donations, and the sale of the tribe's offset credits.

"As we have recovered additional forestlands, we have enrolled additional acreage in California's climate programs in support of our Tribe's strategic goals including protecting salmon habitat, sustaining the revitalization of our cultural lifeways, and facilitating economic self-sufficiency," Mais wrote.

"It's insulting to claim that the Yurok Tribe has 'gamed' or 'exploited' California's climate regulations," he added. "Equally important, it's concerning that elite institutions now criticize us for legally and ethically using a program that was created to protect mature forests and then using those funds to purchase and restore more forest land that was, at one point, ours."

New Forests defended its practices in emailed responses to questions, arguing its projects have preserved existing carbon stocks and removed CO_2 from the atmosphere through subsequent tree growth "as confirmed via third-party verification."

In a statement, the company said it has worked on projects in numerous areas, not just along the program's regional boundaries. The company said its projects "have protected and will enhance carbon storage on hundreds of thousands of acres of forests," adding that one project with the Chugach Alaska Corporation enabled the permanent retirement of a significant portion of the coal reserves in the Bering River Coal Field in southeastern Alaska.

New Forests follows the board's "scientifically-accepted regulations to both the spirit and letter of the program," the company said in a subsequent statement. "New Forests is proud of the forest carbon projects we have developed under California's climate programs—they have generated positive environmental impact and furthered the economic and cultural objectives of the family forest landowners and Native American tribes with whom we have worked."

New Forests didn't respond to numerous additional inquiries, including direct questions about whether it was gaming the rules of the program.

In an emailed response, CarbonPlan stressed that its paper criticizes the design of the program—not the Yurok Tribe or other landowners. Nor does it allege anyone has broken the rules. Its analysis doesn't consider or depend on the intent of any forest owners, who can benefit from flaws in the rules whether they intend to or even know about them.

"We recognize the injustices experienced by the Yurok Tribe, including the seizure of their historical lands by the United States government and its citizens," the nonprofit stated. "We also recognize the Yurok Tribe's legitimate interest in securing resources to repurchase lands that previously belonged to the Tribe and its people."

An Open Secret

Chris Field, an environmental studies professor at Stanford University, was coauthor of a 2017 study that found California's program was helping to prevent emissions on balance by reducing logging. About 64 percent of the thirty-nine projects studied were "being actively logged at or prior to project inception."

Field said the state program is "relatively well-designed to address key issues," but said it can and should be improved.

He added that there are firm limits on the role that offsets can play in California. From now through 2025, state polluters can only buy offsets to cover as much as 4 percent of their carbon emissions; from 2026 to 2030, that ceiling rises to 6 percent.

But those numbers understate the critical role of offsets in California's cap-and-trade program, viewed by some as a model for market-based climate policy.

Under that program, California sells permits that allow certain industries to emit greenhouse gases, with each permit worth one metric ton of CO_2. The state also regularly gives away a certain number of permits to various regulated companies. The total number of permits, called a "cap," declines over time.

Polluters can also purchase permits from other companies with extras to spare, which constitutes the "trade." Or they can buy carbon offset credits, which cost slightly less than permits.

To participate in the offset program, landowners must hire technicians to survey the trees on their land, then take data such as tree type, height, and diameter and plug it into equations to estimate the carbon stored per acre.

Most of the credits are distributed during the initial stages of a project, which can help to repay setup costs. Projects can also earn additional credits over time as the trees grow and absorb CO_2, but those credits accrue slowly, and are dwarfed by the initial credits given to forests with more carbon than the regional average.

The type of forest projects that CarbonPlan analyzed account for 68 percent of all credits issued by the Air Resources Board since the program's launch, far eclipsing other types of offsets like capturing methane from dairy farms or coal mines, CarbonPlan found.

Cap and trade is designed to slash the state's carbon footprint by 236 million tons of CO_2 over the next decade, about a third of the cumulative reductions needed to meet the state's emissions targets over that time.

Barbara Haya, who leads the Berkeley Carbon Trading Project at UC-Berkeley and is a coauthor of the CarbonPlan study, calculated that up to half of those cap-and-trade emissions cuts could come via offsets.

Haya said these cherry-picking practices have been an open secret. The study is "revealing to everyone what a lot of people in the industry understand," she said.

Conservation vs. the Climate

Supporters of forest offsets say no system is perfect, and that focusing solely on the carbon math overlooks the incentives offsets create for protecting forests.

Field said offset systems should balance two goals: ensuring real emissions cuts and creating ways to fund forest conservation. If CarbonPlan's study shows projects are gravitating toward high-carbon forests, then those are exactly the types of trees you'd want to save "if you have a conservation agenda," he said.

Cody Desautel, president of the Intertribal Timber Council, a Portland-based nonprofit consortium of Native tribes, said that off-set programs have provided critical financial flexibility for tribes. They've allowed them to buy back historic land, build needed in-frastructure, create jobs for members, or simply save up money for financial security. But above all, they've created incentives to manage forests in sustainable ways, he said.

"Tribes are very conservation-minded," said Desautel, who is also the natural resources director for Washington's Confederated Tribes of the Colville Reservation, which operate an offset project under California's system. "Their practices are largely based on what's best for the ecosystem, not what makes the most sense eco-nomically. And there's never been any value to that management approach in the past. These carbon projects provide an opportu-nity to value that."

He added, "If there's no value to owning forestland, it probably won't be forestland long into the future."

The Yurok Tribe's offset projects have clearly helped in these sorts of ways, even if they didn't provide the full promised carbon benefit.

The tribe has said it is using the acquired land and funds to restore its old-growth forests, produce traditional foods and basket-weaving materials, create a salmon sanctuary, and improve habi-tat for endangered or culturally important species like the coho salmon, northern spotted owl, black-tailed deer, and Roosevelt elk.

"Our partnership with New Forests will provide the Tribe with the means to boost biodiversity, accelerate watershed restoration, and increase the abundance of important cultural resources like acorns, huckleberry and hundreds of medicinal plants that thrive in a fully functioning forest ecosystem," Thomas P. O'Rourke Sr., then-chairman of the Yurok Tribal Council, said in a statement at the time.

But if the societal goal is preserving forests, it would be simpler and more effective to describe it accurately and fund it directly, said Haya, the UC-Berkeley expert. As soon as these forests get tied up in an offset program, the carbon math does matter, because every additional ton purportedly preserved in trees enables polluters to purchase the right to generate an additional ton of CO_2.

Forest offsets appeal to the public partly because of what academics call "charismatic carbon"—they offer a feel-good story of environmental and social good.

"Any good conservation advocate would tell you there's a desperate need for more funding, and we agree entirely," CarbonPlan's Cullenward said in an email. The "problem isn't that conservation is bad, it's that the system of carbon offsets channels these real needs and sincere hopes into a system that grinds it all up and spits out garbage on the other side."

"The Best Bang for the Buck"

California's Air Resources Board approved the forest offset program's official rules in 2011, after years of discussions with dozens of experts, including government scientists and staff from conservation groups.

In adopting them, the agency relied heavily on Climate Action Reserve, a nonprofit that created programs with voluntary offset credits. The nonprofit, which continues to advise the agency, led an effort to calculate regional carbon averages as part of an initiative to update its voluntary offset rules.

To do so, the nonprofit used data from the U.S. Forest Service, which surveys tens of thousands of forest plots nationwide. The nonprofit grouped data from different tree species and combined data from various geographic zones into larger regional areas called supersections. This simplification allowed the Climate Action Reserve to create a set of common baselines that estimated the amount of carbon stored in typical privately owned forests. The baselines take into account such forest uses as logging.

But the use of these broad averages obscured real differences on the ground. Some industry insiders and researchers began to notice that landowners and developers routinely located their projects in areas where the specific tract of forest differed greatly from the regional averages.

Zack Parisa, chief executive of the carbon offsets company Silvia-Terra, previously consulted for project developers and landowners enrolling forests in California's system. But he said he stopped out of frustration, after seeing the ways it was regularly being gamed, including the cherry-picking techniques CarbonPlan highlighted.

Parisa said he doesn't blame landowners or project developers, who are acting out of rational self-interest.

"If someone shows up and is offering a contract to buy carbon and it doesn't require them to change anything about how they manage the forests, that's free money and they'd be stupid not to take it," he said.

"I'm not hunting for a villain here," Parisa added. "Of course, they look for the best bang for the buck."

In addition to New Forests, other developers also worked on projects where favorable boundaries and forest types boosted the credits that could be earned, according to CarbonPlan. Those include Bluesource and Finite Carbon, which BP purchased a majority stake in late last year. The researchers found that those two developers' projects, taken together, generated up to twenty-four million credits that don't represent actual carbon reductions.

New Forests, Finite Carbon, Bluesource, and other subjects of this article were provided the full study and an accompanying paper describing its methods.

Finite Carbon declined to address detailed questions but stressed that the Air Resources Board and an independent auditor found that their projects were in compliance with the rules.

In a statement, the company said there were "unanswered questions" about the CarbonPlan study's methodology, adding, "however we cannot comment further on it as the underlying raw data is not currently available for public review."

Emily Six, the marketing and communications manager for Bluesource, denied the company had gamed the rules in any way.

In an email, Six said California's program actually undercounts the carbon preserved through projects by not crediting the amount stored in other parts of the forest like soil, shrubs, and foliage. She also stressed that without offsets, some landowners could have chopped down their forests to carbon levels well below the regional average.

"Deliberately overstating climate benefits would run counter to our very purpose for existence," she wrote. "Bluesource exists to improve the world by improving the environment."

The experts who wrote the original offset rules relied on the only national forest data set available, from the U.S. Forest Service's Forest Inventory and Analysis Program, said Constance Best, co-founder of the Pacific Forest Trust. The conservation nonprofit was closely involved in the creation of the early program and participated in it.

Best said it was necessary to create carbon averages for larger regions and forest types because there wasn't enough fine-grained data to ensure accuracy at highly local levels. She disputed Carbon-Plan's claim that its researchers had created a better way of calculating regional averages, since their method required relying on a smaller number of forest plots.

"The reason some super sections are large is to assure the data is more accurate," Best said in an email. "So, their solution creates more problems."

In a separate note, she said: "The paper you shared has a strong editorial bias that undermines its findings and makes me question their data and analysis. It deliberately exaggerates what they present as smoking gun over-credited projects."

In an emailed statement, CarbonPlan acknowledges that using fewer forest plots entails some uncertainty. But the researchers stressed they clearly accounted for it by providing a range of results, and maintained their findings are more accurate because they considered the specific mix of tree species in each project. CarbonPlan also shot back at the allegation of bias: "Having done our work on the basis of extensive public program records, and with fully reproducible methods, data, and code, we are confident that other researchers are capable of judging our paper on its merits."

While the board has updated regional averages based on more recent forest data, critics say efforts to address more fundamental problems have been thwarted.

Researchers and activists also worry about the close ties between the Air Resources Board and the groups that now profit from the program.

For example, whenever a landowner wants to enroll a forest tract in California's program, they open an account at Climate Action Reserve or two other nonprofits that have received the board's blessing to review the documents.

If the project clears the Climate Action Reserve's review and a subsequent audit by the state board, the nonprofit charges 19 cents for every credit issued. For one of the largest projects in the program, for instance, that would have added up to more than $1 million.

It "strikes me as a massive conflict of interest for an organization—whether nonprofit or not—that designed the system to have a financial stake in its operation," David Victor, a professor at the University of California, San Diego, who has closely studied international offset systems, said in an email. (Victor recently coauthored the book *Making Climate Policy Work* with Cullenward.)

"In any other market, putting the market players in charge of key elements of its design would lead to 'hollers'" over the conflicts of interest, Victor said. With the forest offset program, "everyone seems fine or even happy about the arrangement."

Climate Action Reserve didn't respond to multiple requests for comment.

"Too Good to Be True"

Hardy, drought-tolerant softwoods like junipers and pinyon pines dominate in the hot, dry landscape of central New Mexico, with smatterings of taller Douglas firs and spruces in the cooler, higher reaches of the mountains.

But under the initial rules of California's program, those forests were considered to contain no carbon whatsoever.

The error stemmed from the fact that there was no available Forest Service data in that part of New Mexico when the Climate Action Reserve calculated regional averages, said Olaf Kuegler, a Forest Service statistician who provided technical assistance to the nonprofit on the federal database.

Consequently, the Climate Action Reserve set the regional average for an area stretching nearly 34,000 square miles at zero, which meant anyone who owned a few dozen trees could earn carbon credits.

Kuegler said he wasn't aware of the mistake until early or mid-2014, when Air Resources Board employee Barbara Bamberger asked him about it. Bamberger, who leads the board's work on forest offsets, later highlighted the error during an October 2014 webinar on offsets.

During her presentation, Bamberger said the board was updating the regional averages in ways that could lead to major changes in certain areas.

"This may be due to the fact that no data existed for some years in the original span from years 2002 to 2006," she explained. "For example, in New Mexico data wasn't collected until the end of that period."

Almost exactly one year after Bamberger's presentation, New Forests' affiliate filed the paperwork for a nearly 222,000-acre project in New Mexico, stretching across the Mescalero Apache Tribe's nearly half-million-acre reservation about ninety minutes west of Roswell. More than a third of the project's trees were carbon-rich Douglas firs, according to the project's paperwork. Shillinglaw signed the forms.

The erroneously low carbon calculation allowed the developer to claim they could have heavily logged the forest, boosting the amount of credits they could earn.

The project earned 3.7 million credits for its first year, worth more than $50 million.

When the California board's updated rules went into effect two weeks later, it set a far higher regional average for most of the project area. If that standard had been in place earlier, it would have eliminated nearly every credit the project earned, CarbonPlan found. The project generated more ghost credits than any other in the nonprofit's study, based on its more conservative calculations of regional carbon averages.

The Mescalero Apache Tribe's president at the time, Danny Breuninger Sr., said the tribe welcomed the project.

"None of us had heard about the carbon credit program, and in a way it sounded too good to be true," he said. "But it was a great deal. It worked out great for us."

Breuninger referred further questions to the tribe's current president, Gabe Aguilar. Neither Aguilar nor the tribe's attorney, Nelva Cervantes, responded to repeated inquiries.

In a statement, the Air Resources Board said the project met all the requirements of the program at that time. The fact that the board was in the process of developing new regional averages using data that didn't previously exist didn't make the earlier figures "invalid or erroneous," it added.

*

"A Second Wave of Colonization"

Ghost credits matter because they allow other companies to purchase the right to continue emitting real greenhouse gases.

Credits from the Mescalero Apache Tribe's project were sold to PG&E, Chevron, and a company that drills for oil in Kern County, California, according to the latest figures available.

The Yurok Tribe's 7,660-acre project generated credits that were obtained by a variety of energy companies like Calpine, PG&E, and Shell.

Some tribal members are deeply uncomfortable with the idea of selling offsets to companies like this even if they are legitimate, fearing they're effectively profiting from pollution.

The offsets, by definition, allow California companies to continue producing more CO_2 than otherwise allowed—as well as the toxic pollutants like soot and heavy metals that frequently accompany such emissions—often near poor neighborhoods. Communities near refineries, cement kilns, and power plants have frequently opposed offset programs.

Thomas Joseph, an activist and a member of the Hoopa Valley Tribe in California, said offset developers target tribal projects because tribes are in "dire need of revenue" and own vast tracts of mostly intact forest. He said his tribe has resisted multiple pitches from developers. "For us to use this as a means to allow corporations to continue to pollute," he said, goes "against our cultural values." He added, "I see it as a second wave of colonization."

Desautel, the Intertribal Timber Council president, sees it differently. When the issue comes up among tribal members, he explains that polluters under cap and trade need to pay either the state for permission to pollute, or landowners through carbon offsets.

"The check is getting written one way or the other," he said. "It's just a question of where it goes and what's being accomplished with that funding."

SilviaTerra's Parisa said that landowners and project developers will continue to respond to the incentives created in the program, in ways that overstate climate progress, until the program itself changes.

"We need better rules," he said. "Let's make sure the dollars we spend actually change things.

"Forests really can be a part of the solution for the climate, but we haven't gotten it right yet."

SABRINA IMBLER

In the Oceans, the Volume Is Rising as Never Before

FROM *The New York Times*

ALTHOUGH CLOWN FISH are conceived on coral reefs, they spend the first part of their lives as larvae drifting in the open ocean. The fish are not yet orange, striped, or even capable of swimming. They are still plankton, a term that comes from the Greek word for "wanderer," and wander they do, drifting at the mercy of the currents in an oceanic rumspringa.

When the baby clown fish grow big enough to swim against the tide, they high-tail it home. The fish can't see the reef, but they can hear its snapping, grunting, gurgling, popping, and croaking. These noises make up the soundscape of a healthy reef, and larval fish rely on these soundscapes to find their way back to the reefs, where they will spend the rest of their lives—that is, if they can hear them.

But humans—and their ships, seismic surveys, air guns, pile drivers, dynamite fishing, drilling platforms, speedboats, and even surfing—have made the ocean an unbearably noisy place for marine life, according to a sweeping review of the prevalence and intensity of the impacts of anthropogenic ocean noise published in the journal *Science*. The paper, a collaboration among twenty-five authors from across the globe and various fields of marine acoustics, is the largest synthesis of evidence on the effects of oceanic noise pollution.

"They hit the nail on the head," said Kerri Seger, a senior scientist at Applied Ocean Sciences who was not involved with the research. "By the third page, I was like, 'I'm going to send this to my students.'"

Anthropogenic noise often drowns out the natural soundscapes, putting marine life under immense stress. In the case of baby clown fish, the noise can even doom them to wander the seas without direction, unable to find their way home.

"The cycle is broken," said Carlos Duarte, a marine ecologist at the King Abdullah University of Science and Technology in Saudi Arabia and the lead author on the paper. "The soundtrack of home is now hard to hear, and in many cases has disappeared."

Drowning Out the Signals

In the ocean, visual cues disappear after tens of yards, and chemical cues dissipate after hundreds of yards. But sound can travel thousands of miles and link animals across oceanic basins and in darkness, Dr. Duarte said. As a result, many marine species are impeccably adapted to detect and communicate with sound. Dolphins call one another by unique names. Toadfish hum. Bearded seals trill. Whales sing.

Scientists have been aware of underwater anthropogenic noise, and how far it propagates, for around a century, according to Christine Erbe, the director of the Center for Marine Science and Technology at Curtin University in Perth, Australia, and an author on the paper. But early research on how noise might affect marine life focused on how individual large animals responded to temporary noise sources, such as a whale taking a detour around oil rigs during its migration.

The new study maps out how underwater noise affects countless groups of marine life, including zooplankton and jellyfish. "The extent of the problem of noise pollution has only recently dawned on us," Dr. Erbe wrote in an email.

The idea for the paper came to Dr. Duarte seven years ago. He had been aware of the importance of ocean sound for much of his long career as an ecologist, but he felt that the issue was not recognized on a global scale. Dr. Duarte found that the scientific community that focused on ocean soundscapes was relatively small and siloed, with marine mammal vocalizations in one corner, and underwater seismic activity, acoustic tomography, and policy-makers in other, distant corners. "We've all been on our little gold rushes," said Steve Simpson, a marine biologist at the University of Exeter in England and an author on the paper.

Dr. Duarte wanted to bring together the various corners to synthesize all the evidence they had gathered into a single conversation; maybe something this grand would finally result in policy changes.

The authors screened more than ten thousand papers to ensure they captured every tendril of marine acoustics research from the past few decades, according to Dr. Simpson. Patterns quickly emerged demonstrating the detrimental effects that noise has on almost all marine life. "With all that research, you realize you know more than you think you know," he said.

Dr. Simpson has studied underwater bioacoustics—how fish and marine invertebrates perceive their environment and communicate through sound—for twenty years. Out in the field, he became accustomed to waiting for a passing ship to rumble by before going back to work studying the fish. "I realized, 'Oh wait, these fish experience ships coming by every day,'" he said.

Marine life can adapt to noise pollution by swimming, crawling, or oozing away from it, which means some animals are more successful than others. Whales can learn to skirt busy shipping lanes and fish can dodge the thrum of an approaching fishing vessel, but benthic creatures like slow-moving sea cucumbers have little recourse.

If the noise settles in more permanently, some animals simply leave for good. When acoustic harassment devices were installed to deter seals from preying on salmon farms in the Broughton Archipelago in British Columbia, killer whale populations declined significantly until the devices were removed, according to a 2002 study.

These forced evacuations reduce population sizes as more animals give up territory and compete for the same pools of resources. And certain species that are bound to limited biogeographic ranges, such as the endangered Maui dolphin, have nowhere else to go. "Animals can't avoid the sound because it's everywhere," Dr. Duarte said.

Even temporary sounds can cause chronic hearing damage in the sea creatures unlucky enough to be caught in the acoustic wake. Both fish and marine mammals have hair cells, sensory receptors for hearing. Fish can regrow these cells, but marine mammals probably cannot.

Luckily, unlike greenhouse gases or chemicals, sound is a relatively controllable pollutant. "Noise is about the easiest problem to solve in the ocean," Dr. Simpson said. "We know exactly what causes noise, we know where it is, and we know how to stop it."

In Search of Quiet

Many solutions to anthropogenic noise pollution already exist and are even quite simple. "Slow down, move the shipping lane, avoid sensitive areas, change propellers," Dr. Simpson said. Many ships rely on propellers that cause a great deal of cavitation: tiny bubbles form around the propeller blade and produce a horrible screeching noise. But quieter designs exist or are in the works.

"Propeller design is a very fast-moving technological space," Dr. Simpson said. Other innovations include bubble curtains, which can wrap around a pile driver and insulate the sound.

The researchers also flagged deep-sea mining as an emergent industry that could become a major source of underwater noise and suggested that new technologies could be designed to minimize sound before commercial mining starts.

The authors hope the review connects with policy makers, who have historically ignored noise as a significant anthropogenic stressor on marine life. The United Nations Law of the Sea BBNJ agreement, a document that manages biodiversity in areas beyond national jurisdiction, does not mention noise among its list of cumulative impacts.

The UN's fourteenth sustainable development goal, which focuses on underwater life, does not explicitly mention noise, according to Dr. Seger of Applied Ocean Sciences. "The UN had an ocean noise week where they sat down and listened to it and then went on to another topic," she said.

The paper in *Science* went through three rounds of editing, the last of which occurred after COVID-19 had created many unplanned experiments: shipping activity slowed down, the oceans fell relatively silent, and marine mammals and sharks returned to previously noisy waterways, where they were rarely seen. "Recovery can be almost immediate," Dr. Duarte said.

Alive with Sound

A healthy ocean is not a silent ocean—hail crackling into white-crested waves, glaciers thudding into water, gases burbling from hydrothermal vents, and countless creatures chittering, rasping, and singing are all signs of a normal environment. One of the twenty authors on the paper is the multimedia artist Jana Winderen, who created a six-minute audio track that shifts from a healthy ocean—the calls of bearded seals, snapping crustaceans, and rain—to a disturbed ocean, with motorboats and pile driving.

A year ago, while studying invasive species in seagrass meadows in waters near Greece, Dr. Duarte was just about to come up for air when he heard a horrendous rumble above him: "a huge warship on top of me, going at full speed." He stayed glued to the seafloor until the navy vessel passed, careful to slow down his breathing and not deplete his tank. Around ten minutes later, the sound ebbed, and Dr. Duarte was able to come up safely for air. "I have sympathy for these creatures," he said.

When warships and other anthropogenic noises cease, seagrass meadows have a soundscape entirely their own. In the daytime, the photosynthesizing meadows generate tiny bubbles of oxygen that wobble up the water column, growing until they burst. All together, the bubble blasts make a scintillating sound like many little bells, beckoning larval fish to come home.

The Nature of Plastics

FROM *Orion*

EARLY IN 2004, a buoy was released into the waters off Argentina. Half of the buoy was dark and the other light, like a planet in relief. The buoy sailed east, accompanied by the vastness of the ocean and all the life it contains, the long-lived great humpback whales with their complex songs that carry for miles and the short-lived Argentine shortfin squid. Along the way, many thousands of minuscule creatures were colonizing this new surface, which had appeared like a life raft in the open waters of the South Atlantic.

The researchers who'd dropped the buoy followed its movement in hopes of learning more about ocean currents than generations of science and sailing history had revealed. They watched the buoy float into the wide-open ocean between South America and Africa, those twin coastlines that struck me, as I gazed at them on the pull-down map in first grade, as two puzzle pieces that once linked. They surveilled its movements by GPS. Eighteen months later, the signal ceased. Silence from the satellites.

The buoy continued along the currents of the South Atlantic, free from surveillance, sheltered and shocked by sun and clouds and storms overhead. It was likely molded out of a thermoplastic polymer called acrylonitrile butadiene styrene, or ABS, which, like most plastics, was crafted from the extracted remains of long-ago life-forms. ABS was engineered in the lab to endure—rigid, resilient, capable of withstanding all that being let loose at sea may foist upon it.

All plastic begins in a factory. That much we know. But where it goes next remains poorly understood. Only 1 percent of the plastic released into the marine environment is accounted for, found on the surface and in the intestines of aquatic animals. The rest is a little harder to measure. Some presumably washes back ashore. An untold amount settles, sunk by the weight of its new passengers. (One study found four times more plastic fibers in the sediment of the deep-sea floor than on the surface of the ocean.)

And some, like the buoy, just keeps drifting along.

I have spent thirty years fixated on environmental issues, spawned during my own oceanic migration in the fall of 1989. For a semester, I circumnavigated the planet with five hundred other undergraduates on a decaying coal-fired cruise ship held together by layers of paint. We spent half our time exploring ports and half on board, immersed in classes. One course I took, depressingly titled "Environmental Problems," was taught by a dull Russian professor. The year had been a tumultuous one for humans and nation-states around the globe. We were at sea when the Berlin Wall fell. We traveled through the gray streets of Kiev in the dying days of the USSR and within the walled city of Dubrovnik soon before Yugoslavia dissolved into civil war. We boycotted China, after its government opened fire on youth protestors in Tiananmen Square a few months earlier. Everywhere, life was simmering, boiling over.

It should have been politics I latched on to. But I became preoccupied by what the Russian professor was telling us and what my young eyes were witnessing: a paltry harvest in the nets of Taiwanese fisherfolk, the great pyramids at Giza dissolving under the pollution of Cairo, the trash we shipmates generated, which was hurled off the back deck of the SS *Universe* as she plowed ever westward to the next alluring port. All our junk seemingly vanished in our wake, not unlike the disposal method of civilization at large, its logic evident in the phrase "to throw away." But of course, away is always somewhere.

The vast majority of what is now at sea began on land, dumped both deliberately and inadvertently, an estimated 8 million metric tons each year. Plastic factories spill preproduction pellets known as nurdles, feedstock of the plastic production pipeline, easy to transport to other factories and easy to form into . . . anything. The nurdles escape. The objects they're molded into escape. Things get used and discarded. Even when optimistically collected and bound for a "sanitary" landfill, things fall off the backs of trucks or fly away with the wind. Rains flush it all down the sewer. The sewer daylights into a river, and the river travels to the sea.

Along the way, I've come to learn, life takes hold.

In the fall of 1971, a young biologist named Ed Carpenter was just starting his scientific career at the Marine Biological Lab in Woods Hole, Massachusetts, near where I now live. Ed was interested in plants and sea life, and he'd decided to head off into the Atlantic Ocean to study them.

The *Atlantis II* sauntered through the waters of the Sargasso Sea, dragging along a net to collect what floated on the neuston layer, where sea meets sky, skin of three-quarters of the planet. Ed was there to study sargassum, the floating mats of brown seaweed that give the sea its name as they collect in the North Atlantic Gyre, formed by the four great currents that swirl around it, which in turn are formed by winds and the rotation of our planet in space. Ed was looking for living things but kept encountering plastics: pellet-shaped, brittle, weathered, sharp, white, reflective. All kinds. All colors. A few revealed their former land-based purposes in service to humans: parts of a syringe, a cigar holder, a piece of jewelry, a button snap. Algae and jellyfish-like creatures, he observed, clung to their surfaces, just as they do to the natural tendrils of sargassum. Every haul, from the first to the last, over a distance spanning 1,300 kilometers, captured plastics.

The data showed an average of 3,500 pieces of plastic per square kilometer in the Sargasso Sea. The closest land was Bermuda, 240 kilometers away.

Ed's findings were published in the prestigious journal *Science* in 1972. "Increasing production of plastics, combined with present waste-disposal practices," he wrote with his coauthor, Ken Smith, "will undoubtedly lead to increases in the concentration of these particles." It was likely the first mention in the literature of synthetic polymers showing up in marine environments. It was, as careers go, a proud moment for the young scientist.

But Ed got pushback. His superiors questioned why a biologist should care about plastic. And he received an unexpected visit from a Society of the Plastics Industry representative. "I got the vibes, so to speak, that he was not too happy about this paper," Ed recently told Anja Krieger for her podcast *Plastisphere*. Ed would publish only one more paper on plastics before leaving the subject and Woods Hole behind.

Around the same time, also in Woods Hole, the nonprofit Sea Education Association (SEA) formed with the mission of training young people in environmental literacy. Through its SEA Semester Program (the scientific equivalent of the cultural program I'd done), the organization stocked research vessels with students and sent them off into the Atlantic Ocean. From the beginning, they took note of the plastics they saw. But starting in 1986, SEA students and researchers began using the neuston net method that Ed had used, systematically documenting the plastics they found.

On June 29, 2010, SEA brought in a fateful haul. A crew of sailors, students, and scientists aboard the SSV *Corwith Cramer*, SEA's forty-meter brigantine-rigged sailing ship and research vessel, was about halfway between New York City and the Western Sahara. It was a calm, sunny day, and they had neuston nets extended off both the starboard and port sides when they noticed more and more plastic debris floating on the surface. "Windrow after windrow of chunks," one student says in a video documenting the day, "even a five-gallon bucket." Although a typical thirty-minute trawl might bring in a couple hundred stray pieces of plastic, this haul had to be cut short after twenty minutes because the nets were straining under the weight of the trash. A single net collected 23,000 pieces of plastic.

In the video, a young woman in a teal tank top, hair pulled back, takes a filet knife to a triggerfish they pulled in, an out-of-place fish that seems to have found an ecosystem among the junk. She flays it, slicing off an end of the intestine and squeezing out the contents. "This fish was eating at least thirty pieces of plastic," she says, poking at the colorful bits in the petri dish and passing samples of the muscle and liver to a fellow researcher, who wraps them in tinfoil, bound for a lab in Norway that will analyze bioaccumulation of persistent organic pollutants in the flesh.

"The amount of plastic pieces, the concentration, trumps every other plastic tow that SEA has done in the Atlantic for the last twenty-two years," she says later in the video, her face pulled down in a troubling frown, averting her eyes from the camera as she tries to bottle up her emotions. She's so young. Perhaps she was learning to walk around the time SEA began its surveys. Perhaps her parents were courting when Ed Carpenter took to the Sargasso Sea.

Ed's data from less than forty years earlier estimated a square kilometer of the ocean's surface held 3,500 pieces of plastic. Using this singular staggering collection from the SSV *Corwith Cramer* translates to 26 million pieces of plastic—a 740,000 percent increase—drifting along in the same square area of the sea's neuston layer. Although that one haul was an outlier, it—just like record-shattering heat records around the planet—is part of a clear trend, one that tilts ever upward. The mass of plastic is now double that of all animals, terrestrial and marine. More heat, more plastic. Both manifestations of an industrial world extruded from oil.

"A historic day," a crew member says off camera. "It's an historic day," the teal-topped young woman echoes, adding, "unfortunately," as she turns away.

Plastic, wrote Roland Barthes in 1957, "is less an object than the trace of a movement."

The researchers who dropped their buoy off Argentina lost all trace of its signal before long. Ed Carpenter, venturing farther out, found himself pushed back by a wave of interests. SEA might have gotten to the heart of the matter when they pulled in the mother lode, but in the end, their nets couldn't hold the entirety of the truth either. Maybe the young woman on the deck of SEA's ship, filet knife in hand, cut closer to the point: plastics are never just objects; they're bodies migrating within ecosystems and through other, larger bodies. Pull a single strand of some synthetic filament, as John Muir might now remark, and unravel the universe. Plastic threads hitching everything to everything else.

Back in 2004, microbiologist Linda Amaral-Zettler wasn't thinking much about plastics when she was invited to contribute to the Census of Marine Life, a global research project to take stock of the world's aquatic ecosystems. At the time, she was based in Woods Hole, her "scientific birthplace," she says. (She is now at the Royal Netherlands Institute for Sea Research.) Although her work began with the study of the minute microbial manifestations of life, she, too, like Carpenter, found herself encountering plastic—seemingly inert plastic—instead. Or, more accurately, her explorations unveiled how interactions between the two created a third world, something altogether new. In 2013, she dubbed this place the "plastisphere." Linda's husband and research partner, Erik Zettler, who is also a microbiologist, likened it to an exoplanet, an unexplored place with unknown conditions. How will existing life-forms interact with it?

In January 2019, Linda and Erik's investigations led them to the middle of the South Atlantic Ocean, where she spotted, floating on the surface of the sea, a stray buoy, a bit larger than a basketball, half light and half dark. "It looked brand new," Linda told me. "It really brought home the reality that plastic is meant to last a long time, and it really does." Although the tracker was long dead, Linda could read the number 39257 etched into its surface and used it to track down the drifter's origins: the coast of Argentina, fifteen years prior. The buoy's contributions to the original researchers' studies might have been long past, but Linda began her own line of inquiry. In her hand, she held a buoy, yes, but she also held a plastisphere, an unexplored exoplanet made of ABS. She held an animate realm that promised to answer some of the questions that drive her and Erik. How does all this spent human-made trash become home for living creatures? Who inhabits it, and when, and why? What does it mean for them, and what might it mean for us?

Back in her Woods Hole lab, Linda studied the buoy up close, under the gaze of a scanning electron microscope and with the assistance of chemical analysis and gene sequencing techniques. The dark hemisphere of the buoy, the part that had been submerged for fifteen years, interested her the most. It was teeming with life. The buoy's plastisphere revealed gooseneck barnacles, crustaceans that medieval naturalists once believed spawned actual geese, before they figured out that birds migrate. There were colonies of bryozoans, tiny invertebrates that can't quite survive unless clustered with their kind, and hydroids, an early life stage of a jellyfish. She saw nudibranchs, mollusks who shed their shells to become sea slugs, revealing bodies like blown glass in lightning shades of fuchsia, teal, hot pink. On the microscopic level, there was even more. Much more. An entire aquatic commonwealth was there.

In their studies, Linda and Erik have found that members of these communities arrive in a clear pattern of succession. The early colonizers, photosynthetic ones that spin sunlight into chemical energy that transforms into substance, can affix to a piece of plastic within six hours. Then come larger ones, who graze upon the first arrivals, and then more formidable predators, larger still. There are parasitic relationships and symbiotic ones. There are pit formers, spherical cells that seem integrated into the surface of plastics, like beads stitched into fabric. Linda has found a thousand different species on plastic pieces no bigger than the size of an eye on a LEGO figure.

Since Linda first called what she found under her microscope the "plastisphere," the definition of that word has expanded, as has our understanding of what it means to have plastics woven throughout ecological and biological systems great and small. Researchers from all disciplines are asking questions about how plastics, the synthetic polymers that petrochemical companies are wizards at mass-producing, have become integrated into the workings of our natural—and unnatural—world. What role have plastics played in species interactions? How do beetle larvae live for weeks ingesting nothing but polystyrene? How do entire marine ecosystems survive transoceanic journeys on plastics? What are we to make of the fact that plastics are showing up in tap water and air samples? Why do ocean plastics smell so appetizing to turtles? What is the natural and unnatural world we are making for ourselves, and what is the import for every other living being on Earth?

As plastics degrade, so do the barriers that once seemed so defined and distinct, between inert and organic, between outside a body and its interior, between science and art, between present and future.

Even between what is living and what is not.

Although Linda Amaral-Zettler was studying microbial life-forms on the buoy, the object itself was what is known as a macroplastic, an object she could sight from the deck of a ship. But plastics break down. More wind, more weather. More light, more time. They don't disappear. They just fracture, again and again. Smaller than five millimeters, the size of a pencil eraser's nub, and these polymer pieces are considered microplastics. But how small can a piece of plastic get? Below a hundred nanometers in size—the average size of a coronavirus particle, not remotely visible to the human eye—and the polymers step down the scale from microplastic to nanoplastic. Peer into the most powerful microscopes and there are plastics so small they transgress built boundaries, trespass across biological boundaries. The plastisphere becomes "part of our essence," Linda told me, "our fiber." It's as though our lives are one part of the larger story of plastics, rather than the other way around. As though the (plastic) bottle is the message, and we live inside it.

Meanwhile, life moves with plastic and plastic moves with life. The persistence of plastics and their evolving participation in ecological systems are teaching us how those systems work. The 2011 Japanese earthquake and ensuing tsunami created, in the words of marine scientist James T. Carlton of Williams College, "an extraordinary transoceanic biological rafting event with no known historical precedent." He traced the paths of hundreds of Japanese coastal marine species as they voyaged thousands of miles, over years. They fastened themselves onto boats, buoys, crates, and entire docks. They landed as far away as Alaska and California. A 188-ton piece of a former dock washed up in Oregon with a hundred species, not one of them native to the United States. It's yet another way for species endemic to one place to become invasive elsewhere.

Tracey Williams, who walks the beaches near her home in Cornwall, England, crowdsources information about the plastics that wash ashore and tracks them back to known spills. Back in 1997, a rogue wave swatted sixty-two containers off a cargo ship in the North Atlantic, and one of them held nearly five million LEGOs, including many with an oceanic theme, now set free within the true ocean: octopuses and life rafts, scuba tanks and spear guns. It is a living experiment, tracked via Tracey's Twitter account @LegoLostatSea, exploring how long these synthetic seafarers—constructed with the same indestructible ABS plastic as buoys—will continue their nautical adventures. (I imagine the reaction of a hyperintelligent octopus encountering its ersatz doppelgänger drifting along, those eyes latching onto its inanimate imitation, taking its measure . . .)

Like octopuses, humans are visual creatures. It's easy for us to imagine the movement of objects that even children can grasp in their hands, those that catch our eyes on a seaside walk. We can manage images of LEGOs adrift and can mentally, if not emotionally, process horrifying images of Laysan albatross carcasses—feathers, bones, and beaks flared around piles of plastic on Midway

Atoll, thousands of miles from the nearest continent. We fixate on the image of a turtle with a straw impaled in its nostril, a whale's fins entangled with ghost lines from lost fishing gear. The images are Instagrammable. But do they deceive? These are the macroplastics of our world. They represent only a fraction of what's at sea, because even the hardiest plastics break down.

Environmental mariner Captain Charles Moore, sailing the doldrums of the Pacific Ocean in 1997, put it bluntly: "Let it be said straight up that what we came upon [in the Pacific gyre] was not a mountain of trash, an island of trash, a raft of trash, or a swirling vortex of trash—all media-concocted embellishments of the truth," he wrote. "It was and is a thin plastic soup."

And yet the flawed imagery of some monolithic Garbage Patch lives on. It's powerful, imagining a solid heap of oceanic trash. I've interviewed people who dedicated their lives to environmental cleanup after first hearing about it, read about others who are busy inventing the equivalent of marine trash collectors to scoop it all up and save the sea.

Others who look closer see both less and more. To them, even "soup" doesn't seem like the right descriptor for the plastics they find in the sea.

"It's worse. It's everywhere," writes Max Liboiron, a microplastics geographer at Memorial University of Newfoundland and Labrador and a descendant of the Indigenous Michif and the settlers who came to the region. They direct the Civic Laboratory for Environmental Action Research (CLEAR), a feminist, anticolonial marine science laboratory. Back in 2015, they joined a team of other scientists and citizen scientists—surfers and schoolteachers and recycling consultants—collecting microplastics from Bermuda to New York. Together, they pondered the same question: how best to describe what they'd seen? They arrived upon the phrase "plastic smog," riffing off the familiar petrochemical haze we recognize as we gaze, earthbound, at our horizons. Something both everywhere and nowhere. Something that, no matter how small, amasses into something so substantive it can change the essence of a place, of a body.

The idea of the Great Garbage Patch "served a purpose," Kara Lavender Law tells me. She is SEA's research professor of oceanography who has overseen the organization's student-driven plastic surveys for more than thirty years, and I called her up to ask her about scale. After decades of collecting plastics at sea, what does she make of them? How can we think about plastics as the easy-to-understand image of the Great Garbage Patch disintegrates into something more pervasive, invisible, insidious?

I am thinking of the life we don't register when I ask her this question. I'm thinking of what's inside the cylinder of a straw or deep within the folds of tangled fishing line. But Kara shifts our conversation away from the plastisphere, meaning the objects and the life upon it, and brings it back to bodies. "We're really talking about human health," she says, meaning this is bigger than even the ocean, and smaller than what is visible. "There's plastic in your beer," she says, "and your salt."

Plastics are entering aquifers and passing through water filtration systems. A small but global study—150 tap water samples from five continents—found that 83 percent of the samples contained microplastics, from the shore of Lake Victoria in Uganda to Trump Tower in New York City. The United States was the worst of all nations, with only 6 percent unsullied. I think of my reverse osmosis filter, like the ones that proliferate in places like India, as impenetrable. It is not. Nothing is.

To understand what's in the watersheds she studies, Utah State University biogeochemist Janice Brahney turned to the air. What is in the dust that falls with each rain? She crossed the American West, seeking wilderness, Arctic snow, and glaciers to set up stations where she could collect the motes that exist like fairy dust around us. Back at the lab, she found microbeads, tiny plastics found in beauty products, of every color. She found fibers from clothing—nylon, polyester, polypropylene. Dry fleece fibers brought by cross-continental jet streams from laundered metropolises to the most isolated spots, even the aptly named Craters of the Moon National Monument. Janice found that, on average, plastics constitute 4 percent of dust, an enormous amount, she said.

It is not just the presence of plastics that interests scientists, but also the host of compounds secreted inside, remnants of their manufacture and their lives at sea. There are thousands of additives—including colorants, flame retardants, and plasticizers—that make some kinds of plastic more supple, malleable, and useful. Over time, these additives can leak or leach, polluting the environment around them. But plastics function as a sponge, both releasing to and sopping up these environmental pollutants from the waters around them. One study by Japanese geochemist Hideshige Takada found that three-millimeter, preproduction plastic pellets contained persistent organic pollutants such as dioxins, PCBs, and DDT at a million times the concentration of the surrounding seawater. When ingested, many of these pollutants can act on biological systems as endocrine disrupters, skewing development, stealing intellectual potential, altering fertility, instigating metabolic diseases, or interfering with hormone signaling, even at the lowest dose.

And ingestion is easy. The plastisphere, that biofouling film of life that forms on plastics, creates an alchemy that masquerades marine debris as enticing bits of fake food for unsuspecting creatures. Chocolate frosting smeared on a plastic cupcake. It smells good to them, and so they eat the plastic, now laden with pollutants from surrounding waters. Once ingested, the pollutants transfer to their bodies and bioaccumulation begins. From plastic to zooplankton, from zooplankton to fish, and so on.

Again, the scale is staggering: certain plants can draw up plastic-infused dust through their vascular systems, embedding it in the vegetative material that becomes food for others. Synthetic fabric fibers have been found in the digestive tracts of more than a hundred species and can pass through the intestinal walls of crabs and fish. Polystyrene microspheres that were lit up in a lab to be fluorescent were found to accumulate in the guts of mussels. They then migrated to the circulatory system, where they remained for forty-eight days. "This was a game-changer," the ecologist Mark Browne told one journalist. "Up until that point everyone thought that these particles would be ingested and then go right out the other side."

Eventually this everywhere plastic reaches even the purest of bodies. The human fetus swims in an "inland ocean" of one, as biologist and writer Sandra Steingraber once put it. Even in this cushioned, contained world of amniotic waters, researchers have found traces of plastics-making chemicals: bisphenols, phthalates, flame retardants. Pollutants eternally searching for equilibrium. Everyone, everything, seeking its place in the world order.

Stratigraphers are geologists concerned with order and what the layers of the Earth suggest about deep time, and they, too, are asking questions about plastics. Just as there are matters of scale in size, so are there temporal scales. Recognizing that Earth's systems have left the stable eleven-thousand-year stretch known as the Holocene, some stratigraphers are searching for a new "golden spike," the place in the rock record where our descendants can definitively point and say, "Here. *This* shows how it all changed." Paleoecological signposts that mark humanity's impact on the planet are far and wide, but the stratigraphers are honing in on the 1950s—the time of postwar economic boom, the Great Acceleration, more of everything, everywhere. Could the plastics of the brief decades since then become part of an actual epoch, a permanent mark in the geological record? The stratigraphers lean toward the name Anthropocene, the age of humans, but perhaps the Plasticene is a more fitting name. Our plastic progeny is everywhere, after all, from Arctic sea ice to the deepest of sea trenches. Plastics move with wind, water, time. They spread and settle. Big pieces become little pieces, like boulders breaking down and turning into sand that runs the length of a shoreline, a beach a place in motion, stretching across time. What better signature of this new epoch?

Geologist Patricia Corcoran and sculptor Kelly Jazvac sought out Kamilo Beach in Hawai'i after Captain Charles Moore mentioned it was a place of marine debris accumulation and convergence. There they encountered plastiglomerate, amalgamations of plastic and basalt and coral and shells fused together—perhaps by beach fires—into something that appears as ready-made sculpture: an archaeological and geological marker of our age, blue synthetic swirls of identification melted into metamorphic rock. A strange beauty in a broken world.

The epoch makers are veering away from using plastics to define the threshold of the Anthropocene, proposing instead the use of radionuclides, the fingerprints of the atomic age—another thing that can cross boundaries and enter the biological world. But with a nearly invisible marker detectable only with specialized equipment, instead of one that every human alive today has interacted with, has something been stolen from our descendants? Could the fossils we're forming today be an inheritance that even a child—maybe human, maybe something beyond human, living in some long-to-come future—might chance upon, making their own discovery? Moving across a land once known as Hawai'i or

Cornwall, pause, kneel down, and grasp a colorful stone. And only upon close inspection, realize that it is no stone at all, but something fused together at a moment when the being's ancestors were binding the natural and unnatural worlds into one.

The dream of these artifacts reminds me of the words of Amitav Ghosh, writing of extreme weather events of the climate crisis: "They are the mysterious work of our own hands returning to haunt us in unthinkable shapes and forms."

All life now lives, now evolves, now dies with the carbon dioxide already released into the atmosphere, the plastics already unleashed everywhere. And will, for a very long time.

There is a disturbing splendor in the destruction the industrial age has wrought, which makes me think about Linda in her lab, inspecting the buoy, discovering how its darkened hemisphere could hold such a thriving layer of the living, not unlike the surface of the planet we live upon. A place that was once lifeless rock, where oceans emerged, and, from them, single-celled entities and then photosynthetic ones that could spin sunlight into substance. Setting a stage for a home where life can become ever more complicated, right to its humpback whales and inquiring scientists with neuston nets and floating laboratories.

But still I return again and again to that image from space of what this planet looks like from a remove—that "lonely blue marble"—only half of it ever illuminated at any given moment, all of it afloat in a vast and endless sea of darkness, vulnerable, isolated, yet teeming with life.

Linda kept the buoy after her initial studies were done. She couldn't throw it away, for scientific reasons, or maybe because she knew there was no *away*. But in the environment of the lab, everything that had once lived upon it died, species by species. How thin that layer is. How utterly, devastatingly fragile.

Humans Are a Part of Nature

Black Bears, Black Liberation

FROM *The Cleanest Line* (Patagonia)

SHARING STORIES about animals is a nurturing part of countless bedtime routines. Often these stories feature a friendly young
bear or a kind and welcoming bear family, allowing the stories to
be comforting and compelling to children, easing them into a
good night's sleep. Think of Winnie-the-Pooh, who was one of my
favorites as a child, and his forest full of animal friends that have
safe adventures and promote kindness and understanding. Or the
story of "Goldilocks and the Three Bears," a frequent choice by my
five-year-old daughter, who is amused by how forgiving the bears
are of Goldilocks despite her entering their house and eating their
food. Before bed we recount this tale and she inevitably asks me
if I've ever met the Three Bears, and of course I say yes. When it
comes to bedtime, I'm a bear biologist second, a mother first.

Many of these storybooks began life as folktales, shared orally by
those who memorized them, with improvisations and unique details emerging within cultural groups. Although you can find bear-
related folktales throughout the United States, there are some
far-lesser-known but well-documented stories from plantations in
South Carolina and coastal Georgia from the eighteenth and nineteenth centuries, when slavery persisted. Reading these tales today,
one thing clearly stands out: They were more than stories. They
were tools for parents to protect their children.

Instead of being lighthearted and wholesome, these stories come
from a dark, complicated past. Most are suspenseful and haunting,
designed to keep kids awake and laboring into the night rather
than slipping off to sleep. Enslaved parents knew the consequences

of unfinished tasks on the plantation were far more brutal than sleep deprivation.

In these tales, the bear characters frequently appear as an obstacle within the story; the protagonist has to outsmart the bear in order to succeed. In the Gullah oral tradition from the Georgia coast, for example, B'rer (aka Brother) Bear is an aggressor causing problems for the focal character, often B'rer Rabbit. Some historians believe the bear is symbolic of the slave master: a larger and stronger body with weapons like long claws and sharp teeth suggest an imbalanced dominance and control over the other animal characters in the forests. Generally, these folktales end with the bear being overcome by a different animal character—one less powerful, but more virtuous and cunning—getting away from or over on the bear. The bear is infrequently, if ever, the protagonist of the story, and the bear's failures (e.g., getting its crops stolen or becoming stuck in a tree) allow the protagonist character to survive, perhaps a sustaining, motivational fantasy.

Until I began research for my upcoming book about the shared history between black bears and humans, I had no idea about these folktales—then I all but disappeared down a rabbit hole. Although I knew Native American history and culture would be a large and important part of detailing the story of black bears in North America, in the back of my mind I wondered whether the history of my own cultural group—African Americans—was also relevant to understanding human-bear coexistence, without knowing how.

As much as African Americans have a long, dynamic, and resilient history, our relationship to nature and the outdoors also brings up overwhelming oppression, brutality, and injustice. But then it dawned on me: perhaps the physical sites of oppression could be the link I was looking for.

Following my hunch, I dove into narratives housed in university libraries and African American history museums. That's where I discovered the children's stories. Outside of folktales, most documented interactions between enslaved people and black bears took place while hunting bears for meat or to eradicate them as a "safety" measure to protect the plantation.

Since Southern plantations were often cleared from dense, wildlife-abundant forests, assigning wild-game hunting to the enslaved was both a money-saving tool and a method of clearing wildlife, including large carnivores like bears, from areas valuable for human use. During this time, black bears were seen as a threat, as is still the case in many parts of North America today.

Certain enslaved men were assigned the role of the main hunter for the plantation. They were encouraged to go into the surrounding forests to hunt for game meat that could be prepared for the slave owners in the Main House (e.g., venison, wild boar, wild turkeys), or to hunt for meat that could be used as a supplemental food source for the slave community. Some of these men became acclaimed in their communities as highly skilled bear hunters. Their talents likely developed by necessity, as the consequences of coming home empty-handed could include violent punishment by slave masters.

Human and black bear relationships through hunting are fascinating and unique for many reasons. Although many continents have various species of bears, the continent of Africa does not and did not during the era of the Transatlantic Slave Trade. Bears represented an entirely new menace, and enslaved Africans and those born into slavery in North America had to acquire a rapid local-ecology education likely based on trial and perhaps deadly error while executing forced labor in an entirely new, and very wild, landscape. An enslaved Black man becoming an expert in black-bear behavior, ecology, and hunting techniques is an incredible feat and has been an untold story.

Likely, the skills gained by certain enslaved hunters allowed for a level of nutrition, and potentially safety, on Southern plantations that otherwise would not have existed. But beyond the logic of food and protection, the hunting of black bears and other wild animals by enslaved men on the plantation has symbolic and emotional meaning as well. In *Bathed in Blood: Hunting and Mastery in the Old South,* historian Nicolas Proctor explores the relationships between enslaved people and their masters in the context of hunting wild game.

"A meal that came from someone other than the slaveholder provided a dramatic and easily recognizable symbol of agency and power within the slave community," Proctor writes.

Knowing this, I am once again struck by the possibility that an individual's ability to hunt and kill a black bear, and temporarily release a community's dependence on the slave master for sustenance, could have served as a morale booster that allowed people to see another day. And even though as a scientist I avoid personifying wildlife, I cannot help but find myself creating a fantasy where black bears were aware of the bondage and suffering of Black people in these spaces and perhaps sacrificed their own lives to aid in these indirect acts of resistance.

As I dug further into the biographies of prominent black-bear hunters in North American history, I paused when I came across Davy Crockett's name. During the height of the slave trade, Crockett was able to build a brave and heroic public profile partially due to his visibility as a bear hunter in Tennessee. And while most of the literature surrounding his character points to his bear hunting for sport, fur trade, and duty, a few documents acknowledge his frequent use of enslaved men to facilitate the successful bear hunts. In fact, many of the most famous bear hunters in American history learned from, were assisted by, and often depended on the skilled labor of Black outdoorspeople.

In the literature, these men almost entirely go unnamed, yet their skills are often noted. Some of the literature suggests that the world may never know which person was the most skilled of the bear hunters—Crockett, the famed huntsman, or the unpaid laborer, since Crockett was almost always accompanied by a man enslaved to himself or to an acquaintance who was brought along for the purpose of ensuring a successful hunt.

The most difficult part of my research was uncovering the ways that my beloved black bears may have caused harm to enslaved African Americans who ran through and hid in very wild areas on their escape to freedom. In some scenarios, the threat of a black bear's potential aggression would entirely discourage someone from even attempting an escape. And did those escaping along underground railroads encounter bears? Were black bears present as heroes like Harriet Tubman guided people to freedom, watching them go just as black bears have watched me explore the same forests as a free woman? These thoughts deepened my sense of purpose.

A boy named Elias Thomas is told to have run away from the plantation in order to avoid being punished with a whipping, but after another enslaved member of his community warned him about black bears in the forests he'd run through, Elias turned around and faced the painful consequences that awaited him.

Another narrative details an enslaved man's attempted escape to freedom, where he ran on foot during the night and hid to sleep during the day. One of the days early into his journey, he hid in a bear hole in a hollow tree roughly ten feet from the ground. Not knowing it was a black bear's den, the man was startled when the animal returned, and he had to fight the bear from inside the hole in order to save himself. Armed with a small knife, he stabbed the bear from behind as it backed into its nook in the tree, causing it to fall to the ground below and break its neck. Although the man continued to hide in the tree for what is described as "a long time," his fears of encountering more bears in his journey caused him to return to the plantation.

Though rare, I did find a few examples where enslaved bear hunters used their experience to help them escape slavery. Charles Bell, for one, "became well acquainted with the topography of the forest around his owner's Georgia plantation through frequent hunting outings," before he successfully escaped slavery. While navigating forests on the way to free Northern states, he was able to access protein-rich meals because of his hunting expertise.

I'm still searching for more stories of triumph: where African Americans took to the forests on foot with babies strapped to their backs and hope in their hearts, wading through swamps, traversing riverbanks, maneuvering past bears, and yet successfully— miraculously—crossing into states and territories where African Americans were free. (Slavery was not abolished in Northern states until 1804, preventing the ability of African Americans to "escape to freedom" in many areas by heading north.) Although I've struggled to find these stories, I believe in my heart that it was possible many of these men, women, and children shared the darkness of these wild spaces with black bears and lived to talk about it because of their own preparedness, courage, and perhaps even an unspoken understanding between human and animal.

I may never know who my enslaved ancestors were, but I am confident that their bravery and pioneering in the outdoors have directly led to the brave and pioneering work I do today and drives me to fight for racial justice and equity in these spaces. Their fight did little to improve their own lives but has made a world of difference to mine. And that alone is enough to keep me going.

CYNTHIA R. GREENLEE

Finding Freedom in the Natural World

FROM *The New York Times*

WHEN ALEXIS NIKOLE NELSON was a kindergartner, she counted a honeysuckle tree among her most cherished friends.

She named the tree Priscilla, after her great-aunt. "I wasn't especially adept at climbing trees," she told me as we walked through the woods near her home in Columbus, Ohio. "But this tree grew in this curved way that it was perfectly manageable for me to just scamper up, sit in the branches, and snack on some honeysuckle flowers."

One might expect such an endearing origin story from Ms. Nelson, known to her 1.7 million TikTok followers as the Black Forager. An urban adventurer who roams everywhere from Central Park to areas closer to home, the twenty-nine-year-old makes short, exuberant videos about edible finds in the woods. She gathers unripe black walnuts for her version of the spiced Italian liqueur nocino and extols the virtues of milkweed, a favorite of monarch butterflies and the base of Ms. Nelson's recipe for air-fried fritters. And it all started in those early years with her inclination to view trees as kinfolk.

Though there are no definitive statistics, foragers have informally reported an increase in the practice during the pandemic. "There are clearly new folks getting involved in the practice, and it seems to be for a variety of reasons," said Patrick Hurley, professor and chair of Environmental Studies at Ursinus College, speaking of his local community in Philadelphia.

Ms. Nelson represents one part of an increasingly visible community. While many younger Black people didn't grow up going to the woods to "shop," they have learned about lesser-known fruits such as serviceberries and the common cold remedy burdock root through books or the internet.

Whether they're herbalists, Great Migration grandbabies in search of Southern roots, shoppers slashing their food budgets, the only Black kid who went to 4-H camp back in the day, or home cooks who want to dazzle guests with a backyard-berry crostata, they're often contending with conflicted histories of disconnection from the land—and a present in which they don't always find nature a sanctuary.

The idea that Black people just don't do the outdoors developed over time and centuries of dispossession, said Justin Robinson. An ethnobotanist, farmer, and cultural historian in Durham, North Carolina, he rejects the term "foraging" and its practice as anything new to Black Americans and humans in general. He believes the word separates the world into a disturbing cultivated-versus-wild binary that doesn't reflect reality.

"It's just what we do," he said. "It's life!"

Mr. Robinson links his love of the land and his work to the warm childhood years he spent following his two farmer grandfathers and the adult years he spent unconsciously replicating one of their gardens. But he knows that Black American history is also a series of profound land-related ruptures, starting with enslavement and forced agricultural labor on territory inhabited by—and taken from—Native peoples. The slave master's meager rations turned the enslaved into naturalists out of both necessity and opportunity.

Slave narratives abound with references to tapping honey and finding food. In a 1937 Works Progress Administration interview, Charles Grandy of Norfolk, Virginia, spoke of his escape during the Civil War and how he subsisted on wild berries for days. Sharecropping and land loss—by physical and legal violence—followed. By the early twentieth century, more Southern rural Black people were migrating to cities around the nation. Some swore never to look back or till the land again.

As Mr. Robinson said, Black American history is a combination of "hood and country." And Larry Gholston is holding down part of that rural heritage.

Come each May, Mr. Gholston eyes the cattle yard a short distance from his home in Toccoa, Georgia. He's searching for something very specific—and, in its natural form, toxic: Phytolacca americana, the pokeweed plant native to the South and Appalachia. A sixty-eight-year-old retiree and community historian, Mr. Gholston is committed to preserving poke sallit, a dish made from pokeweed. For the past thirty years, he has been handpicking small, tender leaves for the Poke Sallit Festival that he holds every Memorial Day.

He's trying to pass down his knowledge to younger people, including his thirty-five-year-old son, Seth Gholston, who deejays the event while his father cooks: Seth can now easily spot the ten-foot-tall plant.

The festival is meant "to maintain our heritage," said Mr. Gholston. "A lot of Black folk will tell you, 'I don't eat that mess, man.' It has connotations of poorness and rural."

Although pokeweed's leaves, berries, and roots are poisonous to varying degrees, many rural Americans once soaked, boiled, and sautéed their leaves into poke sallit (possibly a derivation of "salad"), akin to collard greens. The toothsome dish can send an eater to the hospital if its toxins aren't neutralized. Few people know how to cook it correctly now, and fewer dare; Mr. Gholston, who perfected his technique by drawing from family tradition, is an exception.

"My mom would wash it, cook it," he explained. "Some relatives would serve it for Sunday meals. Others would take it as kind of a spring tonic. Older people back in the day used to take the berries and make wine. People have taken the stalk and fried it like okra."

His emphasis on Black self-reliance aligns with that of newer generations of Black explorers. I thought about his ingenuity when I met Ms. Nelson in Jeffrey Park, a Columbus estate turned public resource. Ms. Nelson is a virtuoso of the woods. A walking, talking compendium of botanical factoids and zany zingers, she encourages fans with her cheeky-but-serious prayer for foragers, "Don't die!" and her trademark gap-toothed smile.

What you don't see in her videos are how closely she looks at trees before she ever touches them, how gently she plucks their leaves, and how often she doesn't take anything at all.

Two deer darted in front of us as she picked up black walnuts from a downed tree branch. It never hurts to follow and see what they're looking at, she said. But I noticed that the animals were cavorting behind a colossal mansion that backs up to the woods. Thinking of the film *Get Out* and one character's early warning to not be alone in the woods with white people, I asked how comfortable she feels.

"I do like dressing up and wearing full makeup. Because who doesn't want to prance through the woods and feel like a woman fairy? But some of it is definitely about looking super approachable," she said. Hoodies are off the list of her approved foraging apparel, exchanged for staid cardigans, even in the chilly Midwest fall.

Imagining oneself as a wood nymph wearing a bold lip and loud peasant dress doesn't totally ward off unwanted attention. Ms. Nelson noted that she has been stopped semi-frequently by random white people and rangers.

This is a common complaint of Black people exploring in nature. Widely publicized incidents in 2020—a Black birder was falsely accused of threatening a white woman in Central Park, and a Black man was attacked while hiking in Indiana—are extreme examples of the sorts of routine encounters foragers say they face.

Mr. Robinson said he once stopped his car to take a look at a stand of colic weed across the highway; minutes later, law enforcement arrived to investigate a theft. "I don't know if that was made up or not, but I was literally in an open field," he said. "I doubt anyone except biblical thieves are digging holes in a field to hide their goods." A short conversation later, he headed home safely.

Fushcia-Ann Hoover, a hydrologist who published "A Black Girl's Guide to Foraging," forages in her Annapolis, Maryland, neighborhood, where she's well-known and makes a point of taking her sister's adorable Shih Tzu with her. She cited cases in which Black campers were assaulted by white people in the outdoors. "If it's so dangerous or risky, then maybe it just becomes easier to say, 'Oh, that's just not something we do,'" she said. "So, then you don't feel the loss."

Similarly, Lady Danni Morinich, a fifty-seven-year-old former ad salesperson in Philadelphia (her title comes from a tiny parcel of Scottish land that friends gave her as a humorous gift), runs a business selling teas, tinctures, and other products sometimes made with foraged herbs. She doesn't romanticize the fact that she's often the only Black person at a wild-foods meetup, or the possible consequences of carrying a folding knife into the field:

"I tell other folks, 'Sometime, you might not want to take that.' Because you can get killed being Black while walking."

As I followed Ms. Nelson along a winding trail, her eyes darted around the ground, up to the canopy and down again. She pointed out an early pawpaw fruit, gleaming green twenty feet above us. It is one of very few things for which she would willingly tramp through poison ivy, she said.

The others are chicken of the woods and morel mushrooms; she laments she doesn't have the mycological Spidey sense to spot the latter. Her knowledge, though, does run deep. She is able to identify plants by the shape of their leaves, whether their berries are crowned, the smell of their roots.

At another fork in the path, we stopped at a leaning tree. For mushrooms, an ailing tree is pay dirt. Ms. Nelson plucked a few medium-size brownish-peach wood ear mushrooms. I joked that the hue would make a perfect neutral lipstick for us—two Black women scouting the wilds. She scrunched one of them and held it to the side of her face. Folded that way, it did resemble a human ear, gruesomely sliced, Van Gogh–style.

"My partner hates it when I do that," she said, giggling. He wasn't keen on sampling the mushrooms candied in simple syrup either.

Cooking for others is a major motivation for Dr. Hoover, the Maryland scientist. She has used Ms. Nelson's magnolia flower experimentations to enhance a stir-fry (they taste like ginger) and flavored water with lemony wild sorrel. She even figured out how to soak acorns, a necessary part of the flour-making process, in her toilet tank.

Her family and friends sometimes roll their eyes good-naturedly at "Fushcia's projects," but for her, Black freedom is the larger, continuing project.

"There is power in being able to name the things that are around you and knowing what they can be used for—or can't be used for," she said. "I do take a growing feeling of independence from that, especially as a Black person in this country. There's a part of me that kind of rebels in knowing and being able to take things because the way we are told we're not supposed to."

JULIA ROSEN

Humanity Is Flushing Away
One of Life's Essential Elements

FROM *The Atlantic*

IN A FIELD of sugar beets outside Cambridge, England, Simon
Kelly stands above a narrow trench gouged into the rusty earth,
roughly fifteen feet deep and thirty feet long. "Welcome to the
pit," says Kelly, a bespectacled, white-bearded geologist in a straw
hat and khaki shirt. "You're seeing something that hasn't been
seen in a long time."

The rock layers exposed in the trench date back more than one
hundred million years, to when England lay submerged beneath
a warm, shallow sea. Kelly—a researcher at a nonprofit geology
consultancy—specializes in marine fossils of that era. (*"Dicrano-
donta vagans!"* he exclaims when I find a stone pocked with the
impressions of tiny clam-like shells, which he asks to keep.) That's
why he had an excavator dig this trench in 2015, and why he has
spent countless hours since then sifting through its trove of trea-
sures. "Going out to Simon's hole, are you?" Kelly's wife dead-
panned when I picked him up on the morning of my visit.

I had come because "Simon's hole" also contained objects of more
recent historical significance: dull, round pebbles that once helped
feed the United Kingdom. By the 1800s, centuries of cultivation
had sapped Britain's soils of nutrients, including phosphorus—
an essential element for crops. At the time, manure and bones
were common sources of phosphorus, and when the country exhausted its domestic reserves, it looked elsewhere for more.

"Great Britain is like a ghoul, searching the continents," wrote

Justus von Liebig, the German chemist who first identified the critical role of phosphorus in agriculture. "Already in her eagerness for bones, she has turned up the battlefields of Leipzig, of Waterloo, and of the Crimea; already from the catacombs of Sicily she has carried away the skeletons of many successive generations."

Then, in the 1840s, geologists discovered phosphorus-rich stones buried in the fields around Cambridge—the same smooth, coffee-colored rocks welded into the walls of Kelly's trench. "This is what they were after," he says, pointing to a layer of bean-to-buckeye-size lumps.

These nodules were initially believed to be fossilized feces, and became known as *coprolites,* meaning "dung stones." Most turned out to be chunks of mineralized sediments, but that did not diminish their utility as fertilizer.

"In the remains of an extinct animal world, England is to find the means of increasing her wealth in agricultural produce," Liebig wrote. "May her excellent population be thus redeemed from poverty and misery!" And it was.

Over the ensuing decades, workers extracted 2 million tons of coprolites, transforming the fields and fens of southeast England into a warren of pits and trenches that dwarfed Simon's hole. Coprolites were sorted, washed, and transported by buggy, train, and canal barge to processing facilities, where they were milled and treated with acid to make superphosphate—the world's first chemical fertilizer.

The rocks helped Britain boost its food supply and consummate the so-called Second Agricultural Revolution (the first "revolution" being the rise of agrarian civilization). Coprolites and other geologic deposits of phosphorus also raised the tantalizing possibility that humans had at last broken free of an age-old biological constraint. For billions of years, life on Earth had struggled against a stubborn lack of phosphorus. Finally, that was about to change.

Life as we know it is carbon based. But every organism requires other elements, too, including nitrogen and phosphorus. Nitrogen is the basis of all proteins, from enzymes to muscles, and the nucleic acids that encode our genes. Phosphorus forms the scaffolding of DNA, cell membranes, and our skeletons; it's a key element in tooth and bone minerals.

Too little of either nutrient will limit the productivity of organisms, and, by extension, entire ecosystems. On short timescales, nitrogen often runs out first. But that scarcity never lasts long, geologically speaking: The atmosphere—which is about 80 percent nitrogen—represents an almost infinite reservoir. And early in the course of evolution, certain microbes developed ways to convert atmospheric nitrogen into biologically available compounds.

Alas, there is no analogous trick for phosphorus, which comes primarily from the Earth's crust. Organisms have generally had to wait for geologic forces to crush, dissolve, or otherwise abuse the planet's surface until it weeps phosphorus. This process of weathering can take thousands, even millions, of years. And once phosphorus finally enters the ocean or the soil, where organisms might make use of it, a large fraction reacts into inaccessible chemical forms.

For these reasons, the writer and chemist Isaac Asimov, in a 1959 essay, dubbed phosphorus "life's bottleneck." Noah Planavsky, a geochemist at Yale University, says scientists have reached the same conclusion: "It's what really limits the capacity of the biosphere."

One of the lingering mysteries about the origin of life, in fact, is how the earliest organisms got hold of enough phosphorus to assemble their primitive cellular machinery. Some scientists think they must have evolved in environments with abnormally high concentrations of phosphorus, like closed-basin lakes. Others have suggested that bioavailable phosphorus came to Earth in comets or meteorites—a celestial gift that helped kick-start life.

A chronic shortage of phosphorus might also explain why it took so long for oxygen to build up in Earth's atmosphere. Phytoplankton first began belching out the gas about 2.5 billion years ago, with the advent of photosynthesis. But they might not have had enough phosphorus to ramp up production, according to research by Planavsky and others, because the element kept getting bound up in iron minerals in the ocean, helping trap the world in a low-oxygen state for more than a billion years longer.

That we breathe oxygen today—and exist at all—might be thanks to a series of climatic cataclysms that temporarily freed the planet from phosphorus limitation. About seven hundred million years ago, the oceans repeatedly froze over, and glaciers swallowed the continents, chewing up the rock beneath them. When the ice finally thawed, vast quantities of glacial sediment washed into the seas, delivering unprecedented amounts of phosphorus to the simple marine life-forms that then populated the planet.

Planavsky and his colleagues propose that this influx of nutrients gave evolution an opening. Over the next hundred million years or so, the first multicellular animals appeared, and oxygen concentrations finally began to climb toward modern levels. Scientists still debate exactly what happened, but phosphorus likely played a part. (To Planavsky, it's "one of the most fascinating unresolved questions about our planet's history.")

Another group of scientists, led by Jim Elser of Arizona State University, speculate that such a pulse of phosphorus could have had other evolutionary consequences: since too much phosphorus can be harmful, animals might have started building bones as a way of tying up excess nutrients. "Mind-blowing, right?" Elser says. "If true."

What's clear is that after this explosion of life, the phosphorus vise clamped down again. Geologic weathering kept doling out meager rations of the nutrient, and ecosystems developed ways to conserve and recycle it. (In lakes, for instance, a phosphorus atom might get used thousands of times before reaching the sediment, Elser says.) Together, these geologic and biologic phosphorus cycles set the pace and productivity of life. Until modern humans came along.

Over the course of several weeks in 1669, a German alchemist named Hennig Brand boiled away 1,500 gallons of urine in hopes of finding the mythical philosopher's stone. Instead, he ended up with a glowing white substance that he called *phosphorus*, meaning "light bearer." It became the fifteenth element in the periodic table, the incendiary material in matches and bombs, and—thanks to the work of Liebig and others—a key element in fertilizer.

Long before phosphorus was discovered, however, humans had invented clever ways of managing their local supplies, says Dana Cordell, who leads the food-systems research group at the University of Technology Sydney, in Australia. There and in the Americas, for example, Indigenous people managed hunting and foraging grounds with fire, which effectively fertilized the landscape with the biologically available phosphorus in ash, among other benefits. In agrarian societies, farmers learned to use compost and manure to maintain the fertility of their fields. Even domestic pigeons played an important role in biblical times; their poop—containing nutrients foraged far and wide—helped sustain the orchards and gardens of desert cities.

But human waste was perhaps the most prized fertilizer of all. Though we, too, need phosphorus (it accounts for about 1 percent of our body mass), most of the phosphorus we eat passes through us untouched. Depending on diet, about two-thirds of it winds up in urine and the rest in feces. For millennia, people collected these precious substances—often in the wee hours, giving rise to the term *night soil*—and used them to grow food.

The sewage of the Aztec empire fed its famous floating gardens. Excreta became so valuable that authorities in seventeenth-century Edo, Japan, outlawed toilets that emptied into waterways. And in China the industry of collecting night soil became known as "the business of the golden juice." In Shanghai in 1908, a visiting American soil scientist named Franklin Hiram King reported that the "privilege" of gathering 78,000 tons of human by-products cost the equivalent of $31,000.

King, a forefather of the organic-farming movement who briefly worked at the U.S. Department of Agriculture, admired this careful reuse of waste and lamented that he saw nothing like it at home. This, King wrote, was an unfortunate side effect of modern sanitation, which "we esteem one of the great achievements of our civilization."

The so-called Sanitation Revolution followed close on the heels of the Industrial Revolution. In the 1700s and 1800s, Europeans and Americans moved to cities in unprecedented numbers, robbing the land of their waste and the phosphorus therein. This waste soon became an urban scourge, unleashing tides of infectious disease that compelled leaders in places like London to devise ways to shunt away the copious excretions of their residents.

Liebig and other Victorian thinkers argued that this sewage should be transported back to the countryside and sold to farmers as fertilizer. But the volumes involved posed logistical challenges, and critics raised concerns about the safety of sewage farms—as well as their smell. Thus, waste ultimately was sent to rudimentary treatment centers for disposal or, more often, dumped into rivers, lakes, and oceans.

This created what Karl Marx described as the "metabolic rift"— a dangerous disconnect between humans and the soils on which they depend—and effectively sundered the human phosphorus cycle, reshaping its loop into a one-way pipe.

"That single disruption has caused global chaos, you could argue," Cordell says. For one thing, it forced farmers to find new sources of phosphorus to replace the nutrients lost every year to city sewers. To make matters worse, agricultural research in the late 1800s suggested that plants required even more phosphorus than previously thought. And so began a frantic race for fertilizer.

Spain and the United States laid claim to uninhabited islands in the Pacific Ocean, where workers harvested towering accumulations of bird droppings. (Among them was Midway Atoll—later a U.S. naval station.) Back home on American soil, fertilizer companies scoured bat caves for guano and processed the bones of the countless bison slaughtered by hide hunters on the Great Plains.

In the course of these exploits, humans reached across vast distances to secure phosphorus. The discovery of coprolites in British fields allowed humans to reach back in time, too, seizing nutrients from another era and short-circuiting the geologic phosphorus cycle altogether. We saw a way to turn the stubborn trickle into a torrent, and that's exactly what we did.

Until the late 1800s, the "stinking stones" that dotted the fields of South Carolina were considered a nuisance. But as the cost of imported guano soared and the Civil War reshaped Southern agriculture, scientists discovered that these nodules of phosphate rock could be processed into decent fertilizer. By 1870, the first U.S. phosphate mines opened near Charleston and along the coast, tearing up fields, forests, and swamps to reach the bedrock below.

A decade later, geologists discovered even larger deposits in Florida. (To this day, most of the phosphorus on American fields and plates comes from the Southeastern United States.) Other massive formations of phosphate rock have since been identified in the American West, China, the Middle East, and northern Africa.

These deposits became increasingly important in the twentieth century, during the Green Revolution (the third revolution in agriculture, if you're keeping track). Plant breeders developed more productive crops to feed the world and farmers nourished them with nitrogen fertilizer, which became readily available after scientists discovered a way of making it from the nitrogen in air. Now, the main limit to crop growth was phosphorus—and as long as the phosphate mines hummed, that was no limit at all. Between 1950 and 2000, global phosphate-rock production increased sixfold, and helped the human population more than double.

But for as long as scientists have understood the importance of phosphorus, people have worried about running out of it. These fears sparked the fertilizer races of the nineteenth century as well as a series of anxious reports in the twentieth century, including one as early as 1939, after President Franklin D. Roosevelt asked Congress to assess the country's phosphate resources so that "continuous and adequate supplies be insured."

There were also cautionary tales: Large deposits of phosphate rock on the tiny Pacific island of Nauru bolstered Australia and New Zealand's agricultural progress during the twentieth century. But by the 1990s, Nauru's mines had run low, leaving its ten thousand residents destitute and the island in ecological ruins. (In recent years, Nauru has housed a controversial immigrant detention center for Australia.)

These events raised a terrifying possibility: What if the phosphorus floodgates were to suddenly slam shut, relegating humanity once more to the confines of their parochial phosphorus loops? What if our liberation from the geologic phosphorus cycle is only temporary?

In recent years, Cordell has voiced concerns that we are fast consuming our richest and most accessible reserves. U.S. phosphate production has fallen by about 50 percent since 1980, and the country—once the world's largest exporter—has become a net importer. According to some estimates, China, now the leading producer, might have only a few decades of supply left. And under current projections, global production of phosphate rock could start to decline well before the end of the century. This represents an existential threat, Cordell says: "We now have a massive population that is dependent on those phosphorus supplies."

Many experts dispute these dire predictions. They argue that peak phosphorus—like peak oil—is a specter that always seems to recede just before its prophecy is fulfilled. Humans will never extract all of the phosphorus from the Earth's crust, they say, and whenever we have needed more in the past, mining companies have found it. "I don't think anybody really knows how much there is," says Achim Dobermann, the chief scientist at the International Fertilizer Association, an industry group. But Dobermann, whose job involves forecasting phosphorus demand, is confident that "whatever it is is going to last several hundred more years."

Simply extracting more phosphate rock might not solve all of our problems, Cordell says. Already, one in six farmers worldwide can't afford fertilizer, and phosphate prices have started to rise. Due to a tragic quirk of geology, many tropical soils also lock away phosphorus efficiently, forcing farmers to apply more fertilizer than their counterparts in other areas of the world.

The grossly unequal distribution of phosphate-rock resources adds an additional layer of geopolitical complexity. Morocco and its disputed territory, Western Sahara, contain about three-quarters of the world's known reserves of phosphate rock, while India, the nations of the European Union, and many other countries depend largely on phosphorus imports. (In 2014, the EU added phosphate rock to its list of critical raw materials with high supply risk and economic importance.) And as U.S. and Chinese deposits dwindle, the world will increasingly rely on Morocco's mines.

We have already glimpsed how the phosphorus supply chain can go haywire. In 2008, at the height of a global food crisis, the cost of phosphate rock spiked by almost 800 percent before dropping again over the next several months. The causes were numerous: a collapsing global economy, increased imports of phosphorus by India, and decreased exports by China. But the lesson was clear: practically speaking, phosphorus is an undeniably finite resource.

I first heard about the potential for a phosphorus catastrophe a few years later, when a farmer friend mentioned casually that we consume mined phosphorus every day and that those mines are running out. The more I learned, the more fascinated I became by the story, not only because of its surprising and arcane details— eating rocks! mining poop!—but because of its universality.

Phosphorus is a classic natural-resource parable: Humans strain against some kind of scarcity for centuries, then finally find a way to overcome it. We extract more and more of what we need— often in the name of improving the human condition, sometimes transforming society through celebrated revolutions. But eventually, and usually too late, we discover the cost of overextraction. And the cost of breaking the phosphorus cycle is not just looming scarcity, but also rampant pollution. "We have a too-little-too-much problem," says Geneviève Metson, an environmental scientist at

Linköping University in Sweden, "which is what makes this conversation very difficult."

At nearly every stage of its journey from mine to field to toilet, phosphorus seeps into the environment. This leakage has more than doubled the pace of the global phosphorus cycle, devastating water quality around the world. One 2017 study estimated that high phosphorus levels impair watersheds covering roughly 40 percent of Earth's land surface and housing about 90 percent of its people. In more concrete terms, this pollution has a tendency to fill water bodies with slimy, stinking scum.

Too much phosphorus—or nitrogen—jolts aquatic ecosystems long accustomed to modest supplies, Elser says, triggering algal blooms that turn the water green, cloudy, and odorous. The algae not only discourage people from recreating in lakes and rivers (people "like to see their toes," Elser observes) but also can produce toxins that harm wildlife and disrupt drinking-water supplies. And when the algae die, decomposition sucks oxygen out of the water, killing fish and creating devastating dead zones.

Indeed, pollution may be the strongest argument for reducing our dependence on mined phosphorus. "If we take all the phosphorus in the ground and move it into the system—ooh, we're done," Elser says. Some researchers have calculated that unchecked human inputs of phosphorus, combined with climate change, could eventually push much of the ocean into an anoxic state persisting for millennia. "I'm pretty sure we don't want to do that," Elser says, chuckling. Such events have occurred numerous times over Earth's history and are thought to have caused several mass extinctions—for instance, when land plants evolved and sent a pulse of newly weathered phosphorus into the ocean.

The clear consensus among phosphorus experts is that humans must start mending the phosphorus cycle to reduce the environmental damage caused by pollution and to waste less of an increasingly scarce resource. Or, as a button I once saw Elser wear put it, SAVE THE P(EE).

Even industry has gotten on board: Yara, one of the world's largest purveyors of fertilizer, recently announced a partnership with the European waste giant Veolia to recycle phosphorus from agricultural and food waste. Dobermann says that for many companies, sustainability "has increasingly taken over as a priority."

Recycling human waste offers the most direct way of closing the phosphorus loop. A Canadian company called Ostara has installed systems to extract phosphorus from wastewater at municipal treatment plants in more than twenty cities around the world, including Chicago and Atlanta. Switzerland and Germany have even passed laws mandating the recovery of phosphorus from sewage that will take effect over the next decade.

The potential of recapturing phosphorus from animal manure is even greater. "If you get forty Ph.D.s in a room, we always end up talking about cow shit," Elser says. That's because there's a lot of it. And because the last great disruption to the phosphorus cycle involved livestock.

Throughout most of human history, farmers raised crops and animals side by side, which allowed them to easily recycle manure as fertilizer. During the twentieth century, however, agricultural specialization separated livestock operations and grain growers, often by distances too large to transport manure.

This geographic rift effectively severed the last remaining strand of the human phosphorus cycle. And it led to a surplus of phosphorus in areas of intense animal agriculture, exacerbating pollution problems in places like the Chesapeake Bay, the waterways of Wisconsin's dairy country, and Lake Erie. According to a recent study by Metson and others, fifty pounds of phosphorus are released into the environment for every pound of phosphorus consumed in U.S.-raised beef, more than half of which comes from manure. (For wheat, the ratio is roughly two to one.)

In theory, recapturing this phosphorus could make a big difference. Metson and others estimate that the waste of American livestock contains more than enough phosphorus to support the entire U.S. corn crop; another analysis found that recycling all manure could halve global demand for phosphate rock. We have to change our mindset, says Graham MacDonald, Metson's collaborator and an agricultural geographer at McGill University. "These aren't waste streams," he says. "These are resource streams."

One cool day in December, Joe Harrison and I stand six feet apart, wearing masks, in a fenced gravel lot at Washington State University's Puyallup Research and Extension Center. Harrison is a nutrient-management expert at WSU, and we first met at a sustainable phosphorus conference in 2018, where he told me about the mobile recycling unit he was developing to extract phosphorus from manure.

Now, the contraption sits before me: an 18-foot-long metal funnel folded on a flatbed trailer surrounded by green scaffolding, electrical panels, and an assortment of tubes. Over the past several years, Harrison and his colleagues have towed the unit to dozens of dairies across Washington for trial runs. First, a pump sucks liquid manure into huge plastic tanks, where it gets treated with acid. The slurry then flows through a thick hose into the base of the funnel, where it mixes with other chemicals and begins to form struvite, a pearly white phosphorus-bearing mineral (the researchers add some seed crystals beforehand, to help the reaction along). As the manure is processed, the struvite settles to the bottom for collection.

The project is both clever and pragmatic. It seems unlikely that humans will ever go back to growing all of our food locally on diversified, small-scale farms where manure can be recycled the old-fashioned way. ("It's a shitload of work," Dobermann observes.) But technologies like this offer an opportunity to close the phosphorus loop even over vast distances. For instance, Harrison wants to send the struvite harvested at dairies back to the eastern Washington farms that supply them with feed. "Why don't we capture some of this phosphorus in western Washington and ship it back east where the alfalfa's grown?" he says.

Harrison's unit removes up to 62 percent of phosphorus if the manure has been digested by microbes beforehand—an increasingly common practice that also reduces greenhouse-gas emissions—and 39 percent if not. He calculates that a single cow can produce roughly fifty grams of struvite every single day, which means that, in a year, eight animals could provide about enough phosphorus to fertilize an acre of crops.

Struvite is one of several promising phosphorus fertilizers made by recycling human and animal waste. And it has numerous advantages: it's portable; it doesn't contain pathogens and other contaminants common in waste; and, according to Harrison, it works great as fertilizer. "The alfalfa growers—they want it," he says.

Ostara has been testing its struvite, marketed as Crystal Green, for fifteen years, with encouraging results. Their trials have found that, when blended with conventional fertilizer, struvite increases the yields of many crops, including canola and potatoes, Ahren Britton, Ostara's chief technology officer, says. And growers have noticed. "Frankly, the demand for the product has outstripped the amount that we can recover," he says. (Harrison's collaborator on the mobile project, Keith Bowers, has since joined the company, in part to help expand its agricultural operations.)

Ostara's success and Harrison's pilot project prove that on a small scale, at least, it's possible to reconnect the phosphorus cycle. And for wastewater-treatment plants, doing so is economical; under Clean Water Act regulations, they already have to remove excess phosphorus before discharging effluent. But for farmers, most of whom aren't subject to similar rules, phosphorus recovery is just an added cost, according to Jay Gordon, the policy director of the Washington State Dairy Federation. "There's something there," says Gordon, who joined Harrison and me at the research center. But "it's a big damn Rubik's Cube."

Gordon has tried to broker water-quality trading deals in which cities would pay local farmers to reduce runoff, with little success. Earlier this year, he took a different tack: while hosting a tour of the state's dairies for Starbucks executives, Gordon suggested adding phosphorus to the company's new sustainability program. "This is a world and a national food-security issue," Gordon told them. And farmers can be part of the solution. "I would like to see every dairy farmer be a little miniature fertilizer plant," he says. (When contacted, a Starbucks representative could not offer any information on the impact of Gordon's pitch.)

For the moment, however, the mobile recycling unit sits idle. Harrison says farmers in other states had expressed interest in trying the system, but the pandemic brought operations to a halt. And he's retiring in the spring.

Inside his field lab, Harrison shows me a stack of large cardboard cylinders filled with what looks like sand. Each contains the struvite from a single dairy. The project didn't generate enough to supply the commercial alfalfa growers Harrison had in mind. Still, he estimates that there's about a half ton of fertilizer stored in this shed.

"To be a good guy, I oughta find a home for it," he says, but admits that he's already started throwing some away. Gordon, who operates a six-hundred-acre farm, lets out a little cry of surprise. He sold his dairy herd a few years ago and now grows corn, melons, and alfalfa, among other crops. And he perks up at the mention of free phosphorus: "I know just exactly where it can go."

COREY G. JOHNSON, REBECCA WOOLINGTON,
AND ELI MURRAY

Poisoned—Part 1: The Factory

FROM *The Tampa Bay Times*

PLUMES OF DUST, laced with lead, blow across the factory like a sandstorm. The poison hangs so thick in the air, sometimes the only thing visible is the warm, orange glow from the furnace.

Workers, hundreds of them, sweat through twelve-hour shifts at Gopher Resource in Tampa. They extract lead from used car batteries, melt it down, and turn it into blocks of metal to resell.

Eric Autery, forty-three, came to the plant in the summer of 2017 looking for a fresh start. An army vet from Virginia, he dodged bullets and mine explosions in Afghanistan and Iraq but faced new dangers inside Florida's lone lead smelter.

He worked in the furnace department, skimming impurities off the top of gleaming, molten lead. He moved fast in suffocating heat against a steady mist of fumes. He'd feel his respirator slide on his face, the seal separating from his pooling sweat. He'd smell the metallic stench, like old coins, creeping in.

His complexion turned gray. His body felt heavy. His head pounded.

The level of lead in his blood shot up weeks after he started. Co-workers and supervisors told him he needed to wash better before breaks, or after his shift.

But the poison was bound to enter his body. The amount of lead in the air was seven times what Autery's company-issued respirator could handle.

Autery is among hundreds of workers at Gopher who have been exposed to extreme amounts of lead.

They've inhaled it, been burned by it, been covered in it.

And no one has stopped it.

Tampa Bay Times reporters spent eighteen months examining thousands of pages of regulatory reports and company documents, including data tracking the amount of lead in the air and in workers' blood. They interviewed more than eighty current and former workers, twenty of whom shared their medical records.

The following investigative findings will be detailed in a series of stories starting today:

- Gopher exposed workers for years to levels of lead in the air that were hundreds of times higher than the federal limit. At times, the concentration was considered life-threatening. Workers described regular tasks that left them caked with dust, as though they'd been dunked in powdered sugar.

- Eight out of ten workers from 2014 to 2018 had enough lead in their blood to put them at risk of increased blood pressure, kidney dysfunction, or cardiovascular disease. In the past five years, at least fourteen current and former workers have had heart attacks or strokes, some after working in the most contaminated areas of the plant. One employee spent more than three decades around the poison before dying of heart and kidney disease at fifty-six.

- Gopher knew its factory had too much lead dust, but the company disabled ventilation features that captured fumes and moved slowly to fix faulty mechanical systems. Workers were left vulnerable, wearing respirators that couldn't protect them when poison levels spiked. In 2019, one employee faced an air-lead concentration fifteen times beyond what his respirator could guard against.

- Federal rules required that Gopher provide regular checkups, but the company-contracted doctor didn't tell workers their blood-lead levels put them in danger. When employees had health problems that could be tied to lead exposure, he cleared them to work.

- Gopher rewarded employees with bonuses if they kept the amount of lead in their blood down and punished those who couldn't, a practice that alarmed medical experts and ethicists. Workers took desperate measures to strip metals from their bodies, including undergoing dangerous medical procedures. In the most extreme cases, some donated contaminated blood.

- Dust from the plant has been the suspected cause of lead exposure in at least sixteen children—the sons and daughters of employees who unwittingly carried the poison home in their cars or on the soles of their shoes. A baby girl tested so high for the neurotoxin that her pediatrician recommended she be monitored weekly.

- Federal Occupational Safety and Health Administration regulators haven't inspected the factory for lead contamination since 2014 and missed critical problems in previous visits. Even when top regional safety officials ordered increased inspections of lead businesses across the Southeast, no one came to the only place in Florida that produces the metal.

Company officials would not agree to an interview. Gopher's chief operating officer, Eric Robinson, issued a statement to the *Times* and answered some questions in writing.

He said Gopher has cut average employee blood-lead levels in half since acquiring the plant in 2006 and has invested $140 million to make the factory safer. He also said the company devotes thousands of hours a year to safety training.

"Our people and the communities we serve are the most important part of our work, and that is why our overriding core value is to protect people and communities," Robinson said. "We go to significant lengths to keep our employees safe."

In the last decade, more than a third of the lead battery–recycling factories in the United States have gone out of business, including one in South Carolina that shut down this week. Only ten such factories remain. Minnesota-based Gopher Resource owns two of them.

The company, founded seventy-five years ago, generates hundreds of millions in annual revenue, according to one financial analyst. Its clients have included the U.S. military, battery makers, and ammunition suppliers.

More than three hundred people work at the Tampa location. Many are Black or immigrants. Some came to the plant without a diploma or straight from high school, others as they restarted their lives after arrests or time in prison.

The job offered roughly twenty dollars an hour with sizable bonuses—more money than some workers believed their circumstances would allow.

The factory is about six miles east of downtown Tampa, next to a CSX rail yard and a half mile from Kenly Elementary. Its smokestacks tower above the community of small residential homes, auto-repair shops, and places of worship.

Gopher touts green manufacturing that helps keep thirteen

million batteries out of landfills each year. But over the last decade, the plant has been a key reason why Hillsborough has had more adult lead poisoning cases than any other county in Florida, according to health department reports.

Since 2010, the county has recorded more than 2,400 lead poisoning cases among children and adults, surpassing even Miami-Dade County, which has almost twice as many residents.

Lead wreaks havoc on nearly every system in the body. The health effects are so wide-ranging, they can be blamed entirely on other causes.

Gopher workers have no definitive way to identify if any of their health problems were caused by lead. But many medical conditions could be made worse by repeated and prolonged exposure, especially at the levels found inside the plant.

Ten medical and industrial experts told the *Times* that Gopher clearly needed to lower the contamination levels—some so high, they're typically seen only in developing countries.

Dr. Ana Navas-Acien, an expert in heavy metal toxicity at Columbia University, called worker exposures at Gopher "totally unacceptable."

Inside the factory, the sight of dust alone could be unsettling.

Autery, the army vet, spent just over a year at Gopher. He remembered the first time he walked inside.

"What's all this dust here on the ground?" Autery asked the worker who showed him around.

Lead particles.

"What?" Autery responded. "This isn't dirt?"

No, it's lead.

Inside the Dust Storm

Production runs day and night. Dozens of workers clock in at seven A.M. or seven P.M. A tangle of pipes, hissing hoses, and clanking conveyor belts awaits them in a searing heat.

They feed used car batteries into machinery that crushes them, drains the acid, and separates the lead from plastic shells. The lead is scooped with loader trucks and fed into furnaces that burn at around 1,500 degrees. The metal liquefies there.

It's not unusual for water to hit liquid lead, triggering violent explosions that send molten metal flying. Scars from lead splashes are so common workers refer to them as "tattoos" and consider them a rite of passage.

The lead slides down chutes, making its way into kettles, where it glows like lava against the darkened refinery. Workers sprinkle in chemicals to purify it then pour it into molds, branded with the company's name.

Most of the factory isn't air-conditioned, and the furnaces rarely switch off. Firefighters have responded to workers overexposed to chemicals and others who were dizzy, struggling to breathe, or dehydrated.

Some left the plant on stretchers, as their heart raced or consciousness faded.

Kevin Lewis's heart pounded so hard and fast while he worked in the furnace department, the twenty-six-year-old couldn't catch his breath. He was whisked away by ambulance.

Larry Wheeler became disoriented and fainted while working in one of the dustier areas of the plant. An ambulance rushed the thirty-nine-year-old to the hospital, where medical staff told him to limit his exposure to lead.

James Pitts, forty-nine, blacked out with an erratic heart rate as he walked from the locker room to start his maintenance shift. He was taken by paramedics to the hospital.

Robinson, the Gopher executive, declined to answer questions about specific worker exposures or injuries, citing health privacy laws.

All three men had histories of elevated levels of lead in their bodies while working at Gopher.

Poisons are everywhere inside the factory, including sulfur dioxide, and cancer-causing cadmium and arsenic.

Lead is the most prevalent.

OSHA rules require companies to measure the amount of lead in the air by hooking up monitors to workers.

The rules limit worker exposure to an average of fifty micrograms of lead per cubic meter of air over an eight-hour shift. That's roughly equivalent to a pile of lead dust one millimeter wide, long, and tall. About the size of the tip of a ballpoint pen.

In the factory, lead-infused dust blankets the concrete floors. It is piled in corners and coats the cabs of forklifts and loader trucks. Some areas are so dusty and dim, they look like the gray aftermath of a bomb.

The company built a new plant on the property in 2012 and announced it would quadruple production while operating more safely. A sophisticated ventilation system was supposed to capture the dangerous dust. But it has not worked properly, according to interviews and internal studies from 2012, 2013, and 2017.

As a result, lead in the plant's air routinely has been hundreds of times above the federal limit, lab reports show.

The *Times* obtained and analyzed more than three hundred air samples collected by the company from monitors attached to workers from 2007 to 2019. Lead levels exceeded the protection capabilities of the respirators issued to most workers 16 percent of the time plantwide and 26 percent of the time in the furnace department.

Gopher leaders knew lower numbers were achievable. They had to look no further than their other plant in Eagan, Minnesota.

Tampa employees who traveled to Eagan for meetings or training sessions were stunned by what they saw. The floors were so clean, they joked, you could eat off them.

From 2013 to 2014, the average air-lead reading in Tampa's furnace department was six times higher than Eagan's, according to data submitted to Minnesota regulators and other company records.

The highest air reading anywhere inside the Eagan factory was 2,537 micrograms of lead per cubic meter. That's dozens of times above the federal limit but nowhere near Tampa's highest reading. In Tampa, it was 78,729—or more than 1,500 times the federal limit.

In June 2014, a Tampa employee was exposed to 172,655 micrograms of lead per cubic meter while working in the baghouse, where dust gets routed from other parts of the plant. The next year, an air monitor recorded a lead concentration surpassing 200,000.

Those readings were well above the level federal officials consider life-threatening.

Video taken by a worker from the baghouse in 2014 showed dust billowing through a pipe, a gray-brown cloud painting a haze across the workspace. Equipment buzzed and whistled as workers drove small forklift trucks, without windshields.

Workers described pausing their loader trucks in parts of the plant because it became too dusty to see. They tried to clean the floor with push brooms and shovels, only to toss more dust into the air.

By the end of some shifts, the poisonous dust stuck to their sweaty skin like sand.

A Prevalent Poison

Gopher has repeatedly violated OSHA's regulations on air-lead levels and respirators. But the company in recent years hasn't surpassed the federal agency's rules for the maximum amount of lead allowed inside a worker's body.

That's because OSHA permits workers to have as much as sixty micrograms of lead per deciliter of blood, a figure established forty-two years ago. Many health officials say the OSHA standard is out of touch with modern science, which for decades has established health effects from lead at far lower levels.

The Centers for Disease Control and Prevention says blood-lead levels of five micrograms per deciliter and higher count as elevated. But health officials have recognized that damage from lead, like kidney dysfunction, can occur even below that amount.

The *Times* obtained and analyzed blood-lead tests of more than five hundred Gopher employees from 2014 to 2018. Nearly every worker was exposed to enough of the toxic metal to be at risk of serious health problems.

Nine out of every ten workers averaged levels of lead in their blood higher than five micrograms per deciliter.

Eight out of ten workers averaged levels that put them at risk of increased blood pressure, kidney injury, and cardiovascular disease.

In some of the dustiest areas of the plant, workers had the most metal in their blood: four of every ten furnace workers averaged a blood-lead level of at least twenty micrograms per deciliter from 2014 to 2016. That's quadruple the level the CDC considers elevated.

Lead doesn't stay in the blood long. Some of the metal is excreted in urine or settles into tissues. The rest is mistaken by the body for calcium and absorbed into the skeleton.

A single exposure to low or moderate amounts of lead may not cause lasting damage. But chronic exposure compounds with time and can result in irreversible health effects.

The lead collects in larger and larger bone deposits, creating a bank of poison that can reenter the bloodstream and attack the body's organs for decades.

The *Times* shared its findings with ten medical experts. All of them said workers in the plant had blood-lead levels high enough to experience short- and long-term health consequences. They added that lead exposure could exacerbate issues like hypertension or decreased kidney function.

Proving a specific ailment was caused solely by lead exposure is difficult. Diseases often develop because of a combination of risk factors, like age, genetics, and lifestyle.

The *Times* reviewed company medical records of sixteen former workers, who spent from one year to thirty-three years at the plant and left in the last decade. Seven had at least one lab result indicating possible kidney damage. Eleven had blood tests just before they were hired, and all eleven saw the amount of metal in their blood jump within weeks of starting at the factory.

Plantwide, at least fourteen current and former workers had heart attacks, cardiac arrests, or strokes in the last five years, according to interviews and medical records. All were younger than sixty. Three, like Ric Hattan, were under forty-five.

It's rare to have a heart attack at Hattan's age. Fewer than 1 percent of people younger than forty-five have had one, according to federal data.

Hattan, a former maintenance worker, had blood-lead levels hovering in the mid-teens. He described suffering two heart attacks in his early forties, leaving him so afraid of stressing his heart he hesitated to pick up his three-year-old.

"I'm too young to be having a heart attack," Hattan remembered thinking. "I'm too strong."

Breaking Down

Prospere Dumeus started working at the factory in the fall of 1985. It was then a small, family-owned lead smelter called Gulf Coast Lead. He was twenty-three years old, new to Florida from Haiti, and assigned to the furnace department.

The old factory was not fully enclosed. Breezes swept through the work area, cooling the workers and pushing lead dust outside. The plant had a single furnace and produced a fraction of the metal it does today. Workers took off respirators to talk. They'd eat and smoke cigarettes beside the furnace.

Dumeus's formal education had ended in grade school, but as the years passed, he built a vast knowledge of the machinery and its quirks. When something went wrong, Dumeus could diagnose problems better than many mechanics, his coworkers remembered.

Some hazards were obvious. An explosion splattered molten lead on Dumeus in the fall of 1999, burning his left leg and eye. In 2006, hot liquid lead slipped into his boot and scorched his foot.

The burns troubled Dumeus's sister, Madelaine. She implored her brother to quit.

"I'm telling you," she told her brother. "This job is killing you."

But he loved being there. He talked about it with the same adoration as dominoes in the park, fishing trips, and Bob Marley songs. He bought a cottage-style home shaded by thick palms within a mile of the plant.

The longer Dumeus worked around lead, however, the more his body broke down. Medical records and lab tests from the late 1990s show Dumeus consistently had blood-lead levels five, six, seven, and even eight times what is now considered elevated.

His heart problems began around that time, he noted in a medical form that is part of his records. He was in his late thirties.

Over the next decade, he underwent bypass and valve replacement surgeries. He developed leg ulcers and blood clots. His heart strained as it beat.

Several factors put Dumeus at risk of heart problems: hypertension, smoking, too many fats in his blood. He was diagnosed with coronary artery disease, the most common heart condition in America.

It's also the form of heart disease most commonly associated with lead exposure. Medical research has linked lead to cardiovascular effects in people with low levels in their blood, mere fractions of Dumeus's.

The highest concentration of metal in his blood, forty-five micrograms per deciliter, was measured in June 2006, just over a month after Gopher bought the plant. At the time, Dumeus wore a company-issued respirator that covered only the lower half of his face. Weeks later, company data shows, the amount of lead in the air surpassed the mask's protection capability by roughly five times over.

In his early fifties, his lungs had the strength doctors would expect to see in a hundred-year-old man.

The company last measured the metal in his blood in March 2017. Because lead stays in the blood for such a short period, the tests generally show recent exposure and not what has built in the body over time.

The *Times* determined the amount of poison lodged in Dumeus's bones by analyzing 182 blood-lead tests that he took over his career. The calculation estimated a range of lead stored in leg bone, then multiplied the result based on an estimated weight of the skeleton.

The analysis showed how exposures add up. Dumeus's average blood-lead level of twenty-six micrograms per deciliter ballooned to an estimated 420,000 to 840,000 micrograms of lead in his bones.

No amount of lead in bone is considered safe.

Two doctors reviewed the *Times'* analysis and confirmed the findings. Dr. Brian Schwartz, an expert in chronic lead exposure at Johns Hopkins University, said Dumeus's levels could be likened to ingesting a daily pill for years filled with poison.

By the mid-1990s, the neurotoxin had taken a significant hold in his body, the analysis found.

In the winter of 2017, Dumeus worked his last shift.

That March, he underwent open-heart surgery. After months of difficult recovery, he was told by his personal doctor that he could return to work but forbade him from lifting anything heavier than thirty pounds.

In response, Gopher fired him. Dumeus was devastated, his sister said.

Gopher did not answer questions about Dumeus, citing employee privacy.

Less than two months later, in December 2017, his heart stopped at a church service. He lay without a pulse for at least twenty-seven minutes. Paramedics shocked him twice and revived him.

But his brain had been damaged. His mood became flat, his speech limited, his limbs involuntarily jerky.

He was moved to a rehabilitation center in Clearwater. He stopped eating and suffered from seizures. In early 2019, he was taken to the hospital, where he deteriorated. A doctor pronounced Dumeus dead of coronary artery disease, complications from his brain injury, and kidney disease at 7:57 A.M. on February 21, 2019.

He'd lived fifty-six years. For thirty-two of them, he worked at the plant.

Cleared for Work

Federal rules require Gopher to provide employees with regular medical evaluations, and it's Dr. Bruce Bohnker's role to make sure workers can safely do their jobs.

Bohnker is the medical director of a Tampa clinic that Gopher has hired for the past seven years to monitor employee health.

But when workers had ailments that could be caused or made worse by lead, Bohnker didn't note a possible connection or warn them of the consequences, according to a review by the *Times* of medical files for a dozen workers.

In 2016, Bohnker wrote Dumeus a letter describing findings from his exam and noting his history of heart problems.

Bohnker didn't say in his assessment that Dumeus's heart problems and hypertension could make him more vulnerable to poisons. He didn't note a lab result indicating decreased kidney function. He wrote Dumeus had "a long history of working at the plant with no problems."

The *Times* obtained letters Bohnker wrote to six other workers who had hypertension, signs of possible kidney damage, or both.

"I find no areas of concern from this physical examination related to occupational exposures with Gopher," Bohnker wrote to Dumeus and each of the other workers.

Workers described their exams with Bohnker as cursory and said they didn't get explanations of their lab results, including blood-lead levels.

Bohnker, citing doctor-patient confidentiality, declined to answer questions sent to him about any Gopher employees. He also wouldn't answer questions about his role or about the risks posed to workers at the factory.

Doctors interviewed by the *Times* said they would have told the workers with health problems that continued exposure could make matters worse.

Bohnker spent more than three decades as a navy doctor, retiring in 2005. Records show he's certified in occupational, aerospace, and preventive medicine and has no disciplinary history in Florida.

He is a member of the American College of Occupational and Environmental Medicine, and in 2019, served as president of the Florida chapter.

More than a decade ago, the national organization pushed companies and doctors to adopt stricter standards for removing workers with elevated lead levels instead of relying on the outdated OSHA rules. They said workers with two blood-lead tests of twenty micrograms per deciliter or higher or one at thirty should be removed.

When workers whose files the *Times* reviewed had levels that exceeded twenty micrograms per deciliter or even thirty, Bohnker didn't indicate in his medical opinions that their health could be in jeopardy. Instead, he left unchecked a box on the forms next to "in range where adverse health effects may occur."

"Clinically Insignificant"

Occupational physicians like Bohnker have discretion under federal rules to recommend removing workers from lead exposure, regardless of their blood-lead levels, if the doctor deems exposure puts them at too much risk.

That didn't happen for Eric Telemaque.

Telemaque had earned a reputation at the plant as a hard worker, known for putting in long days and extra time to support his children and family.

He came to Florida in the early 1990s from the island of La Gonâve in Haiti. He worked two jobs in Tampa, sleeping three hours a night, before getting hired to break down old batteries at the factory.

When he started in 2006, at age forty, he already had a high blood pressure of 148/90. The amount of lead in his system quickly increased.

Telemaque had trouble navigating the health care system, in part because he mainly spoke Creole and needed an interpreter. His medical records show he had struggled to control his blood pressure, sometimes going long stretches without medication.

During his appointments with Bohnker, over the course of three years, tests showed Telemaque had extremely elevated levels of a protein in his urine indicating possible kidney damage.

On Telemaque's lab results, the protein levels were circled. But Bohnker didn't mention them in his written opinions or letters.

By his December 2015 physical exam, three separate lab tests indicated Telemaque's kidneys could be damaged. He had worked at the plant for nine years.

His blood pressure was 207/136.

"That's the kind of blood pressure that would actually send somebody to the emergency room," said Dr. Howard Hu, a physician and expert in adult lead exposure at the University of Southern California.

Bohnker wrote in exam paperwork that Telemaque was off his blood-pressure medication.

Telemaque's blood-lead level was below the OSHA standard. But Hu and two other occupational physicians told the *Times* that his health problems—elevated protein in urine, hypertension, and decreased kidney function—meant he should not have been around lead and other poisons.

Bohnker cleared him to work.

"As an occupational physician," Hu said, "that's just bad."

In his medical opinion, Bohnker marked Telemaque's lab tests as "clinically insignificant."

But in a letter to Telemaque, Bohnker noted that one abnormal lab result, an elevated waste product in his blood, could be a sign of kidney problems.

He wrote that Telemaque's blood pressure put him at risk of heart disease, kidney disease, and stroke.

"I strongly recommend that you work to better manage your blood pressure," Bohnker wrote to Telemaque in bold and underlined type. "You should have a local physician to follow you if at all possible."

The doctor wrote he had no concerns about lead exposure, however, using the same language he put in letters to Dumeus and other workers.

Telemaque spent seven more months at the factory.

In July 2016, days before his fiftieth birthday, he suffered a stroke, ending his last shift on the locker room floor. He has had at least two strokes since.

Now fifty-four, his gaze is vacant. He sways while trying to stand. Last year, he started wandering outside his Tampa apartment and getting lost. An assisted living facility is now his home.

Desperate and Motivated

Many workers at Gopher viewed the amount of lead in their blood not as a measure of risk but of their standing with the company. That's because Gopher put pressure on workers to keep levels low.

The company, according to internal documents, would terminate probationary employees in their first six months if they couldn't control their blood-lead levels. More seasoned employees were placed on performance plans.

To become a furnace supervisor, Ko Brown said he was required to have a blood-lead level at or below twenty-one micrograms per deciliter. Advancement at the plant was important to Brown, who started in 2011 with a felony record. Finding another job, especially a good-paying one, wouldn't be easy.

"The money I was making was life-altering," Brown said. "I'm rationalizing everything about this company. Not realizing what it's doing to me—don't even care what it's doing to me."

Brown said he went to a clinic for several weeks of intravenous chelation therapy, a process in which heavy metals are stripped from the body and excreted through urine. It can be dangerous because the treatment doesn't differentiate between good metals like iron and bad ones like lead.

"I was going there every day off I had," Brown said.

He got his promotion.

There were other financial incentives.

The company offered bonuses to workers every few months for keeping levels low, internal records show. In 2012, for example, workers received $330 for having a blood-lead level under seventeen micrograms per deciliter. They received $100 for a level under twenty-three; $50 for a level under twenty-seven.

Medical experts said that tying bonuses to the amount of poison in a worker's blood was unethical.

It wrongly shifted responsibility for exposure levels onto workers, instead of the company, said Arthur Caplan, a bioethicist and founder of the ethics division at New York University's School of Medicine.

"You can't go around blaming them for higher exposures," he said. "It's ridiculous. It's absurd. It's unjust."

Worker cleanliness is vital, including washing hands during breaks and showering after shifts to remove lead dust. But the company still has primary responsibility to limit the amount of contamination in the plant, industrial hygienists and doctors said.

In recent years, the company allotted quarterly bonuses based on a blood-lead average across employees, sometimes pitting workers against one another.

Gopher made it easy to know who was putting the bonuses at risk. The company posted the names of workers with high blood-lead levels on lists inside breakrooms.

Confrontations sometimes broke out between employees when one believed his bonus was in jeopardy because another was raising the average.

Gopher did not directly answer questions about bonuses or its culture. Robinson, the chief operating officer, said programs to reduce lead exposure have encouraged lower blood-lead levels at the factory.

Workers have tried all types of remedies to extract metal from their bodies. In the most extreme cases, they donated contaminated blood or platelets.

Three employees told the *Times* they donated blood. A dozen more said the practice was common.

The workers said they believed it could reduce the amount of lead in their blood before their bimonthly tests. Some said they figured the blood banks would tell them if their donation was a problem.

Medical experts said they'd never heard of such a practice and understood the desperation among workers. But they warned that donating contaminated blood was troubling.

Blood banks don't screen for heavy metal toxicity, as they do for certain diseases. That could result in a patient receiving blood with lead during a transfusion. Doctors also said donating contaminated blood wouldn't significantly help the workers lower their blood-lead levels.

Other employees described taking pills, like EDTA, to cleanse their systems.

EDTA tablets are sold as a form of chelation, which is one of the only medical treatments for lead poisoning. Many physicians believe it comes with considerable risks, including potential kidney damage, so it has generally been reserved for those with the highest blood-lead levels.

Federal rules forbid companies from directing employees to use chelation treatment as a means to evade regulatory limits.

"We do not condone and strongly discourage unsafe practices intended to reduce blood-lead levels," Robinson said.

Other workers shared methods less extreme. They took vinegar pills. They focused on eating leafy vegetables. They tried cilantro, vitamins, fruits, probiotics, and prune and pickle juices.

"People had all their ways of getting their blood-leads low," said Wilbert Townsend, a former furnace supervisor. "And I learned the best way to do it was to stay out of the plant."

Bringing It Home

Lead dust left the factory with some workers, on their shoes, cars, or cell phones. It traveled across Tampa and Brandon and Zephyrhills and into their homes, where their children found it.

The *Times* identified sixteen children of plant employees who had lead in their blood, according to interviews and medical records. When workers discussed the blood-lead levels with pediatricians or the health department, they were told dust from the factory was likely the source of the problem, the workers said.

The Department of Health tracks lead poisoning cases across Florida that may come from old paint, ceramics, cosmetics, or other causes. Statewide, from 2010 through 2014, the agency found roughly 175 cases of workers exposing their children, including 18 in Hillsborough County, according to the department's most recent study.

The cases tied to the factory date to the 2000s before Gopher bought the plant. One child ran his fingers along his dad's truck, coated with lead dust, then put his hands in his mouth, said Joe Galant, who served briefly as the safety manager under the previous ownership. Another worker tracked lead dust home on his boots. The poison infiltrated the carpet, where his son would stick his fingers in his mouth as he crawled.

At least thirteen workers have had children with elevated blood-lead levels, the *Times* found.

The young daughter of Altonio Bradshaw, who worked in the furnace department.

The infant son of Larry Wheeler, who worked in the baghouse.

The elementary school–age son of James Pitts, who worked in maintenance.

The most recent instance occurred last year.

Robinson said Gopher is unaware of lead poisoning cases involving the children of its current employees.

Any amount of lead in a child is considered harmful. The health effects could result in stomach pain, headaches, lowered IQ, and slowed growth.

Adam Risher, who worked in the baghouse, learned his oldest daughter, Cheyenne, had lead in her blood in 2014, when she was four.

During a checkup, she had a blood-lead level of sixteen.

His younger children also had lead in their blood, he said. Ayden, who was two, had a blood-lead level of twelve. His baby, Addison, hit thirty-four.

County health officials investigated the source of exposure and identified dust from the factory as the cause, records show.

Addison's pediatrician said her level was so high, she needed to be monitored weekly, Risher said. The infant had more lead in her blood than many of the factory's furnace workers.

"I don't know what to do," he recalled telling the doctor.

Risher's job was dusty. He and other workers manually collected lead dust whenever the automated system crashed.

He thought about how the lead covered his sweaty body at work. He threw away shoes that might have had dust stuck to them. He considered whether washing his socks and boxers at home was a good idea.

He thought about his kids, his wife who stayed home with them, and the need to make overtime on top of an hourly wage in a job that didn't require a college education. He vowed to somehow make sure his children's blood-lead levels came down.

He switched to a department with less dust before leaving Gopher two years ago.

Many workers believed that in the dusty environment, they

couldn't entirely rid themselves of lead. They worried it stuck to their necks or embedded in their hair.

"If you had a blood-lead level," said Brown, the former furnace supervisor, "you were taking it home."

Colin, Brown's son, had lead in his blood since he was an infant. His levels consistently hovered around nine micrograms per deciliter when he turned two and three. During that time, his dad supervised a shift in the furnace department.

Brown's job allowed his family to buy a new, two-story home in a Pasco County subdivision. But his pride was diluted by fear and guilt. He weighed the job's benefit against the danger.

As a baby, Colin developed slowly. He sat up late and didn't babble, never saying mama or dada. As a toddler, he was diagnosed with autism and attention deficit hyperactivity disorder.

Now, at seven, he loves technology and dissecting how it works. His dad said Colin struggles with stomach issues. And in recent months, he started having seizures. Doctors can't say for sure whether lead has been a factor in any of Colin's health problems, but Brown is suspicious.

In 2019, Colin had the lowest amount of lead recorded in his body. Two years after Brown left the factory.

YESSENIA FUNES

Future Moves

FROM *Atmos*

I'M AT THAT AGE where the people around me are doing one of three things: getting married, having babies, or buying a house. The last is what I'm most interested in. I've always dreamed of my own cozy home, but the rapid manifestation of climate change has made me weary. *Where is safe?*, I ask myself. Shockingly, most people I know are moving down South. In fact, disaster-prone areas in the United States are seeing increases in their populations, according to an analysis from Redfin, a real estate research group.

What does that say about the future of climate migration in the United States? It seems that we're ill prepared for this reality. More people are moving into areas "endangered" by climate change, as Redfin put it, which they very well may have to leave again in the future once the crisis reaches a fever pitch. As for me, I want to live where there's the least chance of impact—a place largely safe from the floods, droughts, and wildfires plaguing much of the nation. After all, these threats may only worsen in the future. Whether they do—and by how much—depends on the destiny our leaders choose.

Welcome to The Frontline, where climate migration is close to home. I'm Yessenia Funes, climate editor of *Atmos*. The World Bank published a report on internal climate migration across the globe. It found that 216 million people may migrate within their countries by 2050 without climate action. We often think of migration as a cross-border phenomenon, but plenty of folks move around within borders, too. What we see in the United States is nothing compared to what's to come everywhere else.

Some disasters are quick. They roll in without warning. These are usually hard to miss because of how violent they feel. Think hurricanes and wildfires—the kind of drama that dominates headlines or evening news shows. Other disasters are slow. They take place over months or years. Take sea level rise and water scarcity, for example. These problems are barely visible to the world at large, yet their ripple effects are very real to those feeling them.

So real, in fact, that some people have no choice but to leave home and start anew. This is the reality that's informed a recent report from the World Bank, which projected future internal climate migration patterns throughout the Global South. The report only looks at "slow-onset climate change impacts," such as water scarcity, lower crop productivity, and sea level rise. These factors are enough to force more than 216 million people to move to different parts of their country by 2050. And this doesn't even include North America or some European countries—or those who migrate after a rapid-onset impact.

"People move for many different reasons," said Viviane Clement, report coauthor and senior climate change specialist at the World Bank. "When you're talking about rapid-onset events, it brings to mind the importance of compounding shocks and how those kinds of shocks can add to the vulnerability of communities and livelihoods on top of slow-onset climate change."

In short: these numbers are probably an understatement.

The report uses climate models that rely on the emissions scenarios and development pathways created by the Intergovernmental Panel on Climate Change. They model six regions: Latin America, Sub-Saharan Africa, North Africa, South Asia, East Asia and the Pacific, and Eastern Europe and Central Asia. Every region may see its own patterns, but the numbers are starkest in Sub-Saharan Africa, which may see as many as 85.7 million internal climate migrants. The increasingly dry climate makes it difficult for agriculture to succeed in the region, creating scenarios like the famine in Madagascar and eventual migration to wherever work and food are available.

In some cases, climate-induced migration takes individuals from one danger zone to another. While sea level rise may push some people away from coasts, an issue like water scarcity may encourage others to move into urban areas where water is readily available—even if that's closest to the coast and, therefore, more prone to flooding and storms. This is where climate-resilient urban planning may be key, Clement said. As is development in rural areas to prevent folks from moving into more vulnerable regions due to separate impacts.

"Coastal areas are going to be both in-migration and out-migration hotspots," Clement said. "The actions that can be taken in order to prepare those localities and make them more climate resilient are going to be different depending on the country. Those solutions are going to need to be tailored accordingly."

Many of the places the report outlines are already experiencing a taste of what this reality may look like—despite being among the least responsible for climate change. That doesn't mean this future is set in stone, Clement said. After all, the report is based on models that can always change should human behavior switch up. And that's ultimately what needs to happen if we want to prevent the instability and disruption that can accompany migration like this. Our behavior needs to change—and policy makers need to incentivize it through laws that curb emissions. They also need to include climate change within migration plans. There's enough warming baked into the planet to ensure people may continue to move around for a while.

And yet there's no predestined outcome. The path we take will be a direct result of climate policy (or lack thereof). Adequate action can prevent up to 80 percent of this predicted migration.

"That window of opportunity to act is still open, so we should be doing so with urgency," Clement said. "That concerted action to reduce greenhouse gas emissions and pursuing green, inclusive, and resilient development are going to be critical in order to reduce that scale of internal climate migration."

Presidents and politicians: this one's on you.

RACHEL RAMIREZ

There's a Clear Fix to Helping Black Communities Fight Pollution

FROM Vox

SHARON LAVIGNE HAS LIVED in St. James Parish, Louisiana, a predominantly Black community, all her life. She remembers when the air wasn't covered with thick gray smog, when the water was still safe to drink, when the gardens were productive and fertile.

But now, she says, "we are sick and we are dying."

Lavigne has watched her neighbors die from cancer and suffer from respiratory illnesses. About five years ago, she, too, was diagnosed with pollution-linked autoimmune hepatitis, with tests showing she had aluminum inside her body. The reason for the community's decline in health, environmentalists say, is a burgeoning fossil fuel industry right in their backyards.

Over the past three decades, roughly 150 chemical plants and refineries have been building facilities up and down the eighty-five-mile stretch of the Mississippi River that straddles New Orleans and Baton Rouge, which includes St. James Parish. According to data from the Environmental Protection Agency (EPA), seven out of ten U.S. census tracts with the country's highest cancer risk levels from air pollution are located in this corridor, known as "Cancer Alley."

So, when Lavigne heard that the Taiwanese plastics manufacturer Formosa was going to build a $9.4 billion petrochemical complex just two miles from her home, she retired from her teaching job in 2018 and started the faith-based environmental justice group RISE St. James to fight the new development project.

Formosa's vast 2,400-acre site, currently marked off with fences, sits on two former nineteenth-century sugarcane plantations and a burial ground for the enslaved, which the company failed to disclose until RISE St. James filed a public records request. Still, the Louisiana Department of Environmental Quality approved permits last year for Formosa to build the complex of fourteen plastics plants, despite the company's own models revealing that it could more than double the amount of toxic pollutants in the area and emit more of the carcinogenic chemical ethylene oxide than almost any other facility in the country.

The predominantly Black communities of St. James Parish and the rest of Louisiana's Cancer Alley are not alone in this problem. According to the National Black Environmental Justice Network, Black Americans in nineteen states are 79 percent more likely to live with industrial pollution than white people. Researchers also found that Black people breathe 56 percent more pollution than they cause, whereas white people breathe 17 percent less pollution than they generate.

Lavigne said industries "come to Black communities because they think no one's going to say anything. They think no one is going to fight."

Environmental groups like RISE St. James usually have one ally in their corner when fighting industrial polluters: the National Environmental Policy Act (NEPA), a bedrock law that requires federal agencies to consider the environmental impacts of proposed infrastructure such as the construction of major highways, prison complexes, airports, pipelines, landfills, and refineries. Passed by Congress in 1969, NEPA, followed by the Clean Air and Clean Water acts, was part of a broader plan to protect the environment from any point source of pollution or contamination.

The law is not perfect, though. Since the link between racism and the environment didn't click for many in the late 1960s and '70s, when these environmental laws were created, NEPA's lack of civil rights protections resulted in the further oppression and exclusion of Black communities across the country. Polluting industries would set up shop in marginalized neighborhoods with no regard to the systemic injustice baked into the fabric of the community, and there was little recourse to stop these polluters from doing so.

But with the rise of the environmental justice movement in the late 1970s, Black environmentalists and policy experts began floating the idea of stronger environmental policies that draw from the 1964 Civil Rights Act. The idea was to protect historically disadvantaged neighborhoods from racist policies that could exacerbate a community's social and environmental burdens.

"People often forget the legacies of slavery, of Jim Crow segregation, and out of that chain, laws that were deeply entrenched within the social structure of the Southern environment that worsened our quality of life," said Beverly Wright, the founder and executive director of the Deep South Center for Environmental Justice, who has advised President Joe Biden on environmental justice policies. "That legacy resulted in communities that had been inundated with toxic facilities, impacting our health, the value of the homes where people live, causing them to have higher cancer rates, and to eventually be relocated from within the midst of these facilities."

With a new Democratic administration, activists say now is the time to marry civil rights protections with NEPA. Strengthening NEPA—often called the "Magna Carta" of environmental laws—by invoking the Civil Rights Act would give underserved populations, like St. James Parish, a greater chance of eliminating the legacy pollution that has choked their communities. Adding these protections, without creating an entirely new policy, wouldn't be very complicated for the Biden administration to do. It wouldn't even need the help of Congress.

The Birth of the Environmental Justice Movement Started with Black People

Environmental injustice—the disproportionate harm that low-income communities and communities of color face from both the causes (fossil fuel pollution) and effects (extreme heat and severe flooding) of climate change—has long been a product of systemic racism.

For instance, a 2019 study found that redlining, the government-sanctioned effort to segregate communities of color that began in the 1930s, is a strong indicator of which neighborhoods suffer the most from extreme heat. While white neighborhoods historically received more community investment in clean green spaces that help cool the area, Black neighborhoods were deprived of resources and slotted next to traffic-choked highways and other industrial infrastructure.

Fossil fuel companies exploited this segregation. In places like Mossville, Louisiana, a small, unincorporated town founded by formerly enslaved people in 1790, nearly all its Black residents have been bought out by the South African petrochemical giant Sasol to build a gargantuan chemical complex. A similar scenario played out in the East End neighborhood of Freeport, Texas, labeled as the "Negro District" in the 1930s. Housing, residents, and once-thriving businesses in East End have dwindled, a trend recently accelerated as officials voted to use eminent domain to expand the port's shipping channels to make room for large polluting industrial ships.

Such systemic injustices are as old as America. But there's growing scientific awareness and pushback against these inequities. Environmental lobbying groups had long been overwhelmingly white, focusing more on nature conservation and less on community impact. It wasn't until recently that Big Green groups began to reckon with their racist past. In the wake of last summer's nationwide protests for racial justice, for example, Sierra Club put out a statement that acknowledged the role it played in perpetuating white supremacy in the movement.

Meanwhile, environmental justice pioneers such as Wright and Robert Bullard, a professor of environmental policy at Texas Southern University, have put out academic research on the ties between systemic racism and its environmental impact on vulnerable communities, which has led to more people being educated and involved in making those connections. It wasn't until Lavigne attended a community advocacy organization meeting in 2017, for instance, that she linked what's been happening to other communities' environment and public health to industrial pollution in her backyard.

"Environmental justice is not a footnote anymore; it's a headline," Bullard said. "Over the last four decades working on this, I realized while we've been able to make a lot of changes over the years, there's still a lot of work that still needs to happen—and it needs to happen in warp speed, because we don't have a lot of time since climate change is with us right now."

The modern environmental justice movement is often traced back to 1978, when a private contractor hired by a transformer-manufacturing company discharged a carcinogenic chemical known as polychlorinated biphenyl (PCBs) in fourteen counties in North Carolina. In response, alarmed residents brought a barrage of complaints and lawsuits against the state and involved parties. This litigation led the state to excavate the 31,000 gallons of soil laced with toxic PCBs, but they needed a place to put it. They chose the small Warren County town of Afton—an overwhelmingly Black, rural, and poor community—as a "suitable" home.

Many environmental scientists questioned just how suitable the location was. Warren County's Black community was especially agitated. Afton residents relied on the town's local wells for drinking water, which could be contaminated by this landfill. For six weeks, residents alongside civil rights groups across the country protested the move. Black activists linked their arms and lay on the ground to block the six thousand dump trucks rolling into their backyards, headed for the newly constructed hazardous waste landfill. Hundreds of protesters were arrested.

"The truth of the matter is that the only way that we got communities of color, especially Black folks, involved in the [environmental justice] movement was by making them see that there was a discriminatory aspect or civil rights violation involved," Wright said. "So, when we talk about some communities having cleaner air than others, it's because of discriminatory policies and for certain it goes directly to civil rights."

In court hearings, NEPA was a first line of defense for communities fighting the landfill. But North Carolina courts carved out an exception to requirements that the state prepare an environmental impact statement, claiming that formal compliance with the law was unnecessary. It was clear to community members and activists that the decision was more politically motivated than based in science: majority-white governmental institutions, likely sympathetic to corporate interests, ultimately allowed a Black, poor, rural, and politically powerless community to be home to a toxic landfill.

Though the residents of Warren County lost, and the toxin-laced soil ended up in the landfills, the incident is still studied by environmental researchers as the hallmark of the environmental justice movement. It was the first major environmental disaster in which civil rights groups, environmentalists, and Black residents fought in solidarity against an act of environmental racism, a term unused until much later.

Today there continues to be no shortage of polluting facilities and infrastructure—factories, highways, waste incinerators, and refineries—being built and erasing low-income communities of color, especially Black neighborhoods. Historians and environmental experts say regulatory agencies, industry executives, and politicians believe it is easier to build in these communities, since many cannot afford to hire legal expertise, or do not have the means to fight back.

"These communities who are affected by disparities in air pollution or just toxic contamination, in general, come from a place of feeling that they are being discriminated against or somehow treated differently from white people," Wright said.

Activists call these places "sacrifice zones," but industrial giants have underestimated how much these typically segregated Black communities will fight for clean air and water, even if they have to do it on their own.

For example, in Port Arthur, Texas, its predominantly Black residents are challenging the mammoth oil and gas refineries that dot the port's skyline and cover the area in thick gray smoke. In Philadelphia last year, Black activists led the fight to shut down the largest oil refinery on the East Coast after years of suffering from facility explosions, bad air quality, and pollution-linked asthma and cancer. In Flint, Michigan, the water that the city's predominantly Black residents had been consuming for years has resulted in serious public health and environmental issues, particularly lead poisoning, that government officials both exacerbated and tried to ignore. Residents are still fighting for a large class-action settlement and government accountability today.

Why It's Vital to Marry Two Historical Policies for Environmental Justice Legislation

Although NEPA has long been a vital tool to shield communities from forms of environmental racism, it isn't a foolproof policy. It requires federal agencies to prepare an environmental impact

statement that describes any environmental or public health ram-
ifications that a development project would pose; however, what
a state or federal agency does with these reports is left entirely to
their discretion. This often means that projects often go forth, regardless of community concerns for potential environmental and
public health impacts.

In most cases, the decision-makers behind the NEPA process
also have financial ties to oil and gas lobbyists or the fossil fuel
industry. Additionally, the public comment process required under NEPA raises accessibility questions, since state agencies tend
to hold hearings far from the proposed site and don't actively
reach out to communities for public input, making it difficult for
impacted communities to engage in the review process. These
loopholes then make it easier for agencies to approve permits for
development projects.

What is in communities' favor is that the NEPA process can take
years, allowing room for activist protests. This has been a major
concern for polluting industries, and it was the driving reason behind the Trump administration's decision to slash the mandated
timeline under NEPA review in 2020. Trump's environmental
agenda focused on weakening, rolling back, and dismantling more
than a hundred critical environmental regulations, such as the Endangered Species Act, the Clean Water Act, and NEPA.

His administration's overhaul of the Magna Carta environmental policy included narrowing the scope of the environmental review process, limiting consideration of safer project alternatives,
and scrapping the requirement to evaluate any project's contribution to climate change. While Biden might be able to reverse the
previous administration's new rules, another Republican president
could undo it again through the same process.

That's why, environmental justice advocates say, vulnerable
communities need a stronger policy now.

Bullard said that taking an integrative approach toward strengthening environmental regulations such as NEPA to include a racial
justice framework is the best way to address the historical neglect in
pollution-burdened communities. He and Wright believe the early
steps the Biden administration has taken in centering environmental justice across his climate and economic agendas, as well as
appointing Michael Regan, the first Black man to lead the EPA, creates an urgency to right the wrongs of the previous administration.

"As climate policies get pushed out, what often gets lost is the policy framing of climate, which has historically been more about dealing with just the science—the parts per million, the greenhouse gases—and not, until the last five to ten years, the justice framing or equity framing," Bullard said. "We have to come up with frameworks that would allow the environmental justice part to get lifted up into the climate framework, because it is a racial justice issue."

Some policy experts say that applying Title VI of the 1964 Civil Rights Act—which prohibits federally funded entities from discriminating on the basis of race in their programs, policies, and projects—to NEPA would help dismantle the environmental inequities rooted in systemic racism that communities face. Title VI alone outlaws intentional discrimination, which many activists allege happens whenever an industrial facility sets up shop in marginalized communities like Afton and St. James Parish.

The task of strengthening NEPA under the Biden administration—including adding back the Trump administration's removed regulatory requirements and adding a civil rights protection mandate—would fall under the domain of the White House Council on Environmental Quality (CEQ). The council would need to solicit new rule recommendations to the bedrock policy, propose new regulations, gather public comments, and conduct public hearings, which could take roughly a year. Biden's new offices of environmental justice under the EPA and Justice Department, which he included in his sweeping executive order on climate change, could also look into avenues that could solidify and protect these vital changes.

In drafting the rules for NEPA to include civil rights protections, CEQ can conduct a Title VI or disparate impact analysis to identify the cumulative environmental and health impacts of adding another polluting facility within a community. If the chosen location is an overwhelmingly Black community already inundated with polluting facilities, for instance, then the state agency shouldn't be allowed to approve permits to develop another project since it will only compound the area's underlying environmental and health conditions.

To enact more permanent change than what can be done through executive action, there are several bills floating around Congress. In February 2020, Democratic Reps. Raúl M. Grijalva of Arizona and Donald McEachin of Virginia introduced a comprehensive bill called the Environmental Justice for All Act—which came as a result of community engagement and more than 350 public comments from community members and leaders of the environmental justice movement.

The bill strengthens NEPA and the 1964 Civil Rights Act, and codifies Bill Clinton's long-standing 1994 executive order that directs federal agencies to identify the disproportionate environmental and human health impacts of any federal actions on low-income communities of color. New Jersey senator Cory Booker has also introduced a separate environmental justice bill that includes reinstating giving individuals the right, under the Civil Rights Act, to bring actions against entities engaging in discriminatory practices.

These are effective yet ambitious policies in addressing the environmental harms Black and other marginalized communities often face. But they may take some time for Congress to take up, despite overwhelming support from Democrats and environmental activists. Adding and strengthening NEPA with civil rights protections, though, does not need to go through Congress—just the CEQ review process that Biden could get started on once his nominee to head the CEQ, Brenda Mallory, the first Black woman to lead the office, gets confirmed.

Bullard is hopeful that Biden will carry out his environmental justice promises: "Knowing the history of the environmental justice movement, it's very important to see how the climate-framing in this new administration, and the policies as they get moved out, that they have taken that justice lens," Bullard said.

Lavigne and the rest of the activists in St. James Parish will continue to hold their ground—and to hold the administration accountable. "Industry will continue to sacrifice Black people's lives to make billions of dollars off of our community," Lavigne said. "It's the new form of slavery. I want President Biden to come down here in Cancer Alley to see what we're going through."

To Be a Field of Poppies

FROM *Harper's*

AMIGO BOB CANTISANO didn't think he was going to die. But his wife, Jenifer Bliss, who had cared for him through eight years of throat cancer, could see that his condition had taken a turn for the worse. The tumor in his neck had imploded, leaving an open wound that Jenifer kept clean and dressed. Little by little, it was growing toward his carotid artery. The doctors warned he could bleed out any day.

"I believe your spirit will live on," she told him, "but your body isn't doing so good. It's important to talk about what you want."

Amigo Bob didn't know what should be done with his body. To bury toxic embalming fluid in the earth was out of the question—he was a lifelong environmentalist. Otherwise, he hadn't given the matter much thought. Then Jenifer heard about human composting.

A few months earlier, in May 2020, a Washington State bill legalizing the conversion of human remains into soil, known as natural organic reduction (NOR), had gone into effect. A company called Recompose was due to open the world's first NOR facility that December in Kent, a city just south of Seattle. They named it the Greenhouse. It seemed perfect for Amigo Bob, who had revolutionized the field of organic agriculture—first as a farmer, then as an advocate and consultant—and spent his life building soil and protecting it from the "pesticide mafia."

When he began to accept that the end was near, Amigo Bob called the founder of Recompose, Katrina Spade. He wanted to make sure she knew what she was doing. Compost is the basis of organic farming, so he knew a lot about it—he'd even served as an adviser for a few large composting operations. Katrina explained their process, and he seemed to find her account convincing, but it wasn't until his final moments that he told Jenifer definitively: "This is what I want."

He died the day after Christmas. His loved ones washed and anointed his body and kept vigil at his bedside. "He looked like a king," Jenifer told me. "He was really, really beautiful." She showed me a few photos. His body had been laid atop a hemp shroud and covered from the neck down in a layer of dried herbs and flower petals. Bouquets of lavender and tree fronds wreathed his head, and a ladybug pendant on a beaded string lay across his brow like a diadem. Only his bearded face was exposed, wearing the peaceful, inscrutable expression of the dead. He did look like a king, or like a woodland deity out of Celtic mythology—his gauze-wrapped neck the only evidence of his life as a mortal.

On the third day of their vigil, Jenifer felt his spirit go.

Amigo Bob joined nine other pioneers at the Greenhouse on the cusp of the New Year: the first humans in the world to be legally composted. Reading their obituaries, I learned that they were as old as ninety-two and as young as forty-eight. One was an "accidental florist," one a "voracious reader," another a "skilled baker" and "serious cook." There was a landscaper, a painter and woodworker, a beekeeper and dog trainer. One taught creative writing to homeless youth, one had a thirty-year career in law enforcement. One man, Ernie Brooks, helped to establish the field of underwater photography and was known as the Ansel Adams of the sea.

Each of their bodies was placed inside an eight-foot-long steel cylinder called a "vessel," along with wood chips, alfalfa, and straw. Over the next thirty days, the Recompose staff monitored the moisture, heat, and pH levels inside the vessels, occasionally rotating them, until the bodies transformed into soil. The soil was then transferred to curing bins, where it remained for two weeks before being tested for toxins and cleared for pickup.

Half of the NOR soil would wind up in a forest on Bells Mountain, in southwestern Washington, near the Oregon border. A composted body produces approximately one cubic yard of soil, which can fill a truck bed and weigh upward of 1,500 pounds. For many surviving relatives—apartment dwellers, for example—taking home such a large quantity of soil is unrealistic, so Recompose offers them the option to donate it to the mountain, where it's used to fertilize trees and repair land degraded by logging.

But Amigo Bob was a farmer, so Jenifer rented a U-Haul and brought the whole cubic yard of him home. She turned the trip into a kind of pilgrimage, stopping to visit loved ones and the headwaters of their favorite rivers. Over the next few months, their farmer friends came by and filled small containers with the soil to use on their own land. Jenifer used some to plant a cherry tree.

I asked her what it was like to have her husband home again, piled up in her driveway.

"Well, it's compost," she told me. "It's still precious because it was his body. But it's also compost."

In my life I have encountered two kinds of people: those who spend time thinking about, talking about, and making plans for their future corpse; and those who prefer not to. I belong to the former category. As a child, I desperately wanted a Viking burial, an idea inspired by the 1988 Macaulay Culkin film *Rocket Gibraltar,* in which a group of kids boost their grandpa's corpse, load it onto a boat, push it out to sea, and light it on fire with a flaming arrow. If the sky glowed red, the narrator explained, it meant the dead Viking had "led a good life."

By my twenties, I had settled on the more realistic option of cremation. I wanted my ashes scattered on the banks of my favorite river, or cast from a cliff into the Pacific Ocean, or fired into the atmosphere from a cannon. (I was in a Hunter S. Thompson phase.) But after a friend's ashes were lost in the mail, I reconsidered. I explored sky burial, in which a corpse is left out in the open to be fed upon by raptors; and alkaline hydrolysis, a process in which flesh is liquefied in a solution of water and potassium hydroxide. More recently, I planned to follow the example of nineties heart-throb Luke Perry and purchase an Infinity Burial Suit: a shroud containing fungi that would consume my corpse and bioremediate its toxins.

Like Jenifer Bliss, I think it's important to talk about what we want, mainly so that our survivors don't have to guess. But I'm also drawn to the death meditation itself, which can lead to useful reflections on life. The appeal of a Viking burial, for instance, was twofold. There was the beauty and drama of uniting the elements in death; I wanted to float, burning over the ocean, and to have the sky take measure of my life. But I also wanted to live like the kids in the movie; kids who build bonfires, shoot bows and arrows, and defy the authorities. In other words, thinking about the kind of death I wanted taught me about the kind of life I wanted.

A willingness to face life's nonnegotiable realities seems to me one mark of psychological maturation. But it comes at a price— the discovery that the world is not as simple as we once believed. Truth contaminates the dream. The Viking burial, for example, is apocryphal; the Vikings were known to burn their dead in boats but kept them parked on land. What's more, their funerals some- times involved human sacrifice, in which a female slave was raped by the dead man's clan, then ritually stabbed and strangled. Other, less sinister realities: both sky burial and the firing of heavy artil- lery are frowned upon in the city of Seattle, where I live. And even if cannons were permitted, cremation releases about 540 pounds of carbon per incinerated corpse. The carbon output from a year's worth of cremations in the United States is roughly equivalent to that from burning 400 million pounds of coal. Alkaline hydrolysis has less ecological impact, but like cremation, it wastes the body's energy; instead of going up in smoke, nutrients are flushed down the drain. Even the mushroom suit, according to critics, adds nothing to the decomposition process that soil itself can't provide. At the end of such a litany, one is liable to conclude, as Dorothy Parker did, "You might as well live."

Of course, that isn't a realistic option either. And so, I approach the dark wood of the middle of my life intrigued to encounter hu- man composting, a method of final disposition with no apparent downside, a method purported to prevent a metric ton of carbon dioxide per body from entering the atmosphere, and to produce soil capable of fertilizing trees and flowers. Whether these benefits withstand the stress of extended consideration remains to be seen. But to leave behind a net-positive legacy, to grow something beauti- ful in death, would be a dream. As a series of attractive promotional cards printed on recycled stock informs me: "I could be a pine- cone," "I could be a forest grove," "I could be a field of poppies."

Katrina Spade, the founder of Recompose, is an architect by train- ing. The idea for human composting first came to her in 2011, when she was a graduate student at UMass Amherst. More pre- cisely, she was drinking a beer in her backyard, watching one of her babies roll around on the grass and marveling at how quickly they grow. It occurred to her that she, too, was growing quickly. But toward what? Oh, right, her certain demise. And what would become of her body when she died?

In retrospect, she now views this revelation as "a little bit trite," but it inspired her to look for alternatives to cremation and burial, neither of which appealed to her. That's when a friend introduced her to livestock mortality composting, a little-known agricultural practice in which farmers inter their expired livestock in outdoor compost piles.

Katrina's master's thesis, "Of Dirt and Decomposition: Proposing a Resting Place for the Urban Dead," contained the seeds of what would eventually become Recompose. She conceived of human composting as a solution to the problem of overcrowded cemeteries and the environmental costs of conventional burial and cremation. But her proposal was also a critique of the premise underpinning those methods—that the body is a disease vector to be disposed of, rather than a potential source of new life. She imagined transforming human remains into soil, "ready to nourish new living beings."

In 2014, Katrina was awarded a fellowship to pursue her idea, and the first stories about her vision for human composting began to appear in the press. Around that time, she received an email from Tanya Marsh, a professor at the Wake Forest University School of Law, who, Katrina recalled, said something like, "Hi, I'm Tanya, I wrote the book on funeral law in the United States, and I just want to let you know what you're proposing to do is illegal. I'd love to talk to you about it." The following semester, Marsh's students began studying human composting in class, trying to figure out which states had the most promising regulatory pathways. It seemed possible that some states' funeral codes might allow for it, but in the end everyone decided that pursuing legalization specific to NOR would be a more effective strategy.

First, Katrina would need to conduct a pilot study with human bodies. She moved her family to Seattle, where she contacted Lynne Carpenter-Boggs, a soil scientist who'd authored studies on livestock mortality composting. In 2018, Katrina and Lynne began a "closed vessel" study with six donor bodies, using a composting drum that was originally devised for livestock remains. An outdoor compost pile is subject to the caprices of weather, and breaking down bones can take more than a year. A closed vessel, on the other hand, would speed up the process by allowing for more control over oxygen, heat, and moisture levels. "You are dramatically ramping up microbial activity," Lynne explained. "You're creating an environment that promotes extremely high activity and heat production."

Composting isn't rocket science, but the process requires a precise amount of sustained heat to eliminate pathogens and quickly convert decaying organic matter to soil. At lower temperatures, "you'll have de-emanation and denaturation," Lynne said, but not true composting. In this sense, the scabby pile of coffee grounds and cut weeds in my yard is actually a decomposition pile. "In the backyard setting," Lynne cautioned, "we do not recommend that people even compost their cat."

The pilot study delivered. Pathogens were eliminated, and pharmaceuticals were remediated to levels well under EPA limits. The closed-vessel system also accelerated the proliferation of the thermophilic organisms that break a body down, transforming it into soil in just thirty days. In a "green burial," by comparison, in which a body is buried in an unlined grave in a shroud or a simple wooden coffin, the process can take up to twenty years.

In the early days, Katrina called her idea the Urban Death Project. It was as direct a name as she could come up with, a way to refuse euphemism in an industry otherwise saturated with it. But it didn't quite capture the regenerative aspect of NOR. So, when she formed her company in 2017, she named it Recompose. The term is canny branding, but it's also a fair description of the process, in that the very molecules of the dead are taken apart and reassembled, as the pilot study put it, into a material that is "unrecognizable visually, chemically, or microbiologically as human remains."

Katrina and Lynne had proved that their process worked, but the legality of NOR was still murky. Katrina had been doing outreach, giving talks, and strategizing about legalization with a local lobbyist. Then, as luck would have it, she realized that a state senator, Jamie Pedersen, lived just down the street. She asked him to coffee and explained what she was up to. "Climate change is high on his list, and he knew his constituents were going to be excited about the idea," she told me. In 2019, Pedersen introduced SB 5001 ("Concerning Human Remains"), the first bill in the country to propose legalizing human composting. "We had six people donate their own bodies to the study before they died—that was their last gesture, and that said something," Katrina told me. "Some of their friends and family testified to the legislature saying, 'This was really meaningful for my person.'" Governor Jay Inslee signed the bill into law that year. (Since then, similar bills have been introduced in California, Vermont, Delaware, Hawaii, Maine, New York, Oregon, and Colorado. The latter two have already passed.)

Meanwhile, Recompose had gained a large following on social media. Its mailing list grew to fifteen thousand subscribers. People all over the world were interested in having their bodies composted. The company had originally leased an 18,500-square-foot warehouse in Seattle's SoDo neighborhood with the intention of installing one hundred vessels. Then the pandemic disrupted funding. An acquaintance offered a deal on the much smaller warehouse space in Kent, and Katrina had to settle for just ten vessels at launch. "My biggest fear was that I'd talk about it for ten years and never do it," she said. The vessels were booked immediately and a wait list began to form.

Evidence of formal hominid burial dates back 120,000 years. Across the ancient world, people interred their dead in large mounds known as tumuli, landforms between three and ninety feet tall, sometimes built in geometric patterns or in the likenesses of animals. Evidence of sky burial dating back twelve thousand years was found at the Neolithic site of Göbekli Tepe in Turkey, but the practice is likely far older. (We don't know exactly how long human beings have practiced sky burial, or cremation, or burial at sea, because evidence of the dead vanishes in the process.) The intentional preservation of corpses through mummification was practiced by the Chinchorro of Chile's Atacama Desert as early as 5000 B.C.E., and has been practiced elsewhere in South America, Asia, the Canary Islands, and of course in Egypt, where priests preserved the dead with oils, plant extracts, and pine resin.

Early American families tended to care for the dead on their own, preparing, dressing, and laying bodies out for viewing in basic wooden coffins. The dead were then buried in the local churchyard or in family plots on the back forty of the farm. The professional undertaker and his industry didn't emerge until the Civil War, alongside the increasingly common practice of embalming, which stalled decomposition long enough for fallen soldiers to travel great distances home for burial. The procedure was both unregulated and profitable, fetching as much as a hundred dollars a corpse. Itinerant embalmers began to trail the Union Army, hovering at the edge of battlefields like kettles of vultures. The most famous embalmed corpse of the time, that of President Lincoln, was a national attraction that passed through seven states before coming to rest in Springfield, Illinois, more than two weeks after his death.

By the 1950s, embalming had become standard in the United States, but I wonder if this would have been the case had people understood the violence involved. There is no single method, but in a typical scenario, fluid containing formaldehyde is pumped into the carotid artery, which forces blood and other fluids in the corpse out of a tube in the jugular or femoral vein. An aspirating device resembling a meat thermometer is then repeatedly pushed into the abdomen and chest, where it punctures the organs. The organs are then filled with concentrated "cavity chemicals." No wonder embalming is considered desecration in some traditions, including among Muslims and Jews, who bury their dead in shrouds or simple coffins, sometimes without nails or fasteners, to avoid obstructing the decomposition process.

What constitutes desecration of a corpse is culture-bound; one man's desecration is another's honorable final disposition. For some, cremation is the only way to release a body's spirit. For others, the idea of burning a loved one's corpse before sending his bones through a pulverizer is the height of barbarism. Ditto the notion of leaving a loved one's remains on a scaffold to be picked apart and consumed by birds, though for the Zoroastrians, who practice sky burial, burying or cremating a corpse would dishonor the sacred elements of earth and fire. Before the practice was condemned by Parliament in the early nineteenth century, people who died by suicide in England were given a profane burial at crossroads. Among early American Puritans, an honorable burial meant orienting the dead's feet toward the east, so they'd rise to face Jerusalem when Christ returned, and Muslims are buried facing Mecca.

The only characteristic that funerary mores seem to share is intentionality. Disposing of the dead in an arbitrary manner— leaving a body where it fell on the battlefield, or tossing it with others into a mass grave, limbs akimbo—is a universal sign of disrespect. Intention is how we signal care, whether or not we believe that the soul persists, or whether we believe in a soul at all.

Surprisingly, burial customs are rarely rooted in earthly practicalities like public-health concerns. Save for a few infectious diseases that remain active in corpses, dead bodies are not generally dangerous. The traditional six-foot burial depth, it turns out, is unnecessary. It's said to have originated during the Black Plague, when people mistakenly believed corpses were the cause of its spread, rather than flea-ridden rats. Decomposition brings with it gases and odors and scavengers, which can be disturbing and unpleasant for the living, but putrefaction itself is not a source of disease. In emergencies that result in mass death, the World Health Organization prioritizes allocating resources to survivors ahead of burying the dead. Our concern, the group says, should be for the living. By my lights, this is also the most convincing argument for being composted.

I visited the Greenhouse one gray afternoon in April, a couple of weeks after Jenifer picked up Amigo Bob. It's a modest warehouse, surrounded by office parks and machine shops. Through the open bay door, I could see the "array": a white wall of interlocking individual vessels. Each vessel rests within a hexagonal frame, so that stacked together they resemble a big white honeycomb. Or some kind of Scandinavian storage solution. Or something out of a seventies sci-fi film: *the galactic travel chambers of the future!* This is not just idle comparison. The clean design is a kind of hedge against inborn anxieties related to dirt and decay, though it does also invite nervous allusions to *Soylent Green.* (Recompose discourages the planting of food crops with NOR soil. Not for any scientific reason, but because the idea makes people uneasy.)

Katrina Spade greeted me at the threshold. We'd only ever talked on the phone, and she was wearing a mask, but I had no trouble identifying her. Whereas many others in the alternative death industry style themselves on a witchy continuum of piercings and botanical tattoos, Katrina wears her hair high and tight, and dresses in trousers and button-downs. Her vibe is warm, but sober, so that when she tells one of her handful of death jokes—"Turns out it doesn't really make good business sense to sell someone a piece of land for eternity. Whose idea was that?"—the listener is disarmed. In other words, there is nothing especially woo-woo about her.

That afternoon, the array was brightly lit by two auxiliary lights. A camera was trained on a white wall, which was staged with a dozen potted plants. Beyond the array, out of sight, the warehouse rang with activity, reinforcing my impression of being on a Hollywood back lot.

"What's up with the stage lights?" I asked Katrina.

She explained that they were preparing for a virtual "laying-in ceremony" that would take place that afternoon. I wasn't allowed to attend. In fact, because of COVID-19 restrictions, they almost didn't let me visit the facility at all. But even in non-COVID times, the warehouse is short on bathrooms and not really set up for public visitation, another reason they plan on moving to a larger facility. For now, if families want a ceremony, they have to do it over Zoom.

Katrina called some of the staff over for introductions. Morgan Yarborough, who previously worked as a funeral director in more conventional settings, manages most of the family logistics. She's also the in-house officiant and would be performing the ceremony later that day. (Families can submit their own words and music, or they can request that Morgan perform something called the carbon cycle ceremony. A representative excerpt: "The plants we are using today, wood chips and straw, will cover Darby's body, powering her transformation and releasing her molecules back into the world.")

Megan Circle is in charge of the soil and the vessels. Her surname—like several others in this story—seems to bear the mark of predestination; in this case, the ashes-to-ashes sense of circularity. As noted in her staff bio: Megan is "the very first person in the world to be employed to usher bodies through their transformation into soil."

Many of these details I already knew. I'd scoped out the website that morning and learned colorful factoids about each. I knew, for example, that in Megan's former life, she worked in the wine industry, trained people in soil regeneration methods, and managed large-scale kitchens. I knew that when she wasn't busy officiating funerals, Morgan kept rescue animals and made pen-and-ink drawings. I knew that the operations manager, Todd Maxfield-Matsumoto, used to work in bookstores, had a background in machine fabrication, founded a record label, and built droids in his spare time as a member of the R2 Builders Club. If human composting attracted a type, I suppose this was it: they all seemed to have a lot of hobbies.

A few days earlier, Megan told me, the team had transported the first five cubic yards of donated NOR soil to Bells Mountain. She loaded their trailer with one yard of compost, but when it came time to load the next, she hesitated. It was the first time the compost would be commingled, which was profound but nerve-racking. "There's a finality to it," she told me. It marked the dissolution of the individual and a return to the collective: "sort of the opposite of a tombstone."

On Bells Mountain, they off-loaded the pile in a clearing in the woods. Morgan Yarborough recited the five names of the dead. Then the living each took up a ceremonial handful of soil and placed it at the base of an alder tree. This wasn't only symbolic of completing a single cycle of renewal, they explained. Alders are pioneer species and are often the first to colonize a clear-cut, fixing nitrogen in the soil and eventually becoming the source of the life that succeeds them.

All the while, as I listened to the staff talk beside the array, the dead were there inside their vessels. And all the while, as I listened, I was listening to something else, too: a kind of ur-tone, the room sound underneath the music of our conversation. Later, when Morgan described the array of vessels as a "hotel for the dead," someone joked that it would make a good band name, and I laughed along. Someone else mentioned a magazine venture: *Better Funeral Homes and Gardens.*

I was hearing and seeing Katrina, and Morgan, and the others. But it was as though I were *remembering* them simultaneously, remembering their voices, and the biographies I had read that morning—texts that bore the same condensed and eclectic color as an obituary.

Oh right, our certain demise. That was the sound beneath the song.

It was, of course, impossible to write this essay without reflecting on the dead people I love, and what I know of their final dispositions. Some were buried, some were cremated and scattered. Others were not so lucky. Last I heard, my grandparents' ashes were stored in my aunt's closet next to her cremated pets, still packed in the cardboard boxes they'd arrived in. Sean's ashes, as I mentioned, were lost in transit. Poor Matt was first embalmed, *then* cremated, *then* buried in his urn. I did not want to be lost in the mail; or mutilated, burned, and buried; or held hostage for decades in my child's closet. I began to feel anxious about my eventual fate, and one morning, after a restless night of battling death in my dreams, I got on the computer and signed up for Precompose, the program that allows you to pay off your future composting in installments as low as twenty-five dollars a month.

Most Americans are squeamish when it comes to death, at least when it comes to considering the prospect for ourselves. This aversion to the realities of our mortal bodies might be a corollary to other historic virtues—vigor, youth, an insatiable lust for the new—but it has had the bizarre effect of stunting innovation in a consumer market that includes literally everyone on the planet. If we broach the subject at all, we do so obliquely, ideally in ways that preserve the option of avoiding death entirely—the apogee of this denial manifest in the cult of anti-aging, the promise of cryostasis and reanimation, in rumors of Walt Disney's frozen head.

My local funeral home pitches grieving families on embalming and heavy-duty caskets as a way to protect corpses from the elements, from the "odors or other unpleasantness that accompany uncared for remains." Such claims are common in the conventional death industry. But the notion that the dead require our protection from decomposition is a fantasy. With few exceptions—such as the continuously maintained corpse of Vladimir Lenin (going strong since 1924)—embalmed bodies break down, too. They just take longer to do so. And rather than contributing nutrients to the earth, they release carcinogens. It seems to me that the promise of protection depends on an unconscious agreement between surviving loved ones and undertakers to play make-believe. To pretend that death need not have the final word. That though we feel helpless, we are not. That we might keep our dead intact, that they are not beyond our care.

I thought paying my Precompose bill every month could serve as a kind of memento mori—a way of resisting death denial. Countless cultural traditions have supplied the living with reminders of mortality, from the baroque bone churches of Europe to the smoke hanging over the Ganges. Theravada Buddhists in Thailand meditate beside corpses as they decompose—all the while reminding themselves: "My body also has this nature." Our poet ancestors had their refrain, *timor mortis conturbat me.* But what does the aging, religiously noncommittal American have? The point of keeping death in mind isn't to dwell on the macabre. The point is to remember what we are always in danger of forgetting: life ends.

I called up a fellow Precompose customer, a seventy-six-year-old practicing psychoanalyst named Linda Wolf, and floated my memento mori idea. She was unmoved. For Linda, it had been a practical consideration, one less thing for her survivors to deal with. She said she hadn't been very conscious of her carbon footprint throughout her life. She knew she owed "the Earth back on that one," and planned to donate her soil to Bells Mountain. It didn't matter to her whether her loved ones had a funeral service or not. "I'm not going to be controlling things from the grave," she said. "I'll be busy fertilizing trees."

"By donating your soil," Recompose tells us, "you have the chance to be productive one last time, providing biomass and nutrients to a forest that truly needs them." Productivity in death might be a selling point for some, but for me (and for others, I suspect) the main appeal of this new method of disposition—which is, in a way, the oldest on Earth—is the opportunity to assuage our guilt and anxiety about the ecological cost of our lives. A process through which mortal fear, both for one's own fate and the fate of the planet, might be sublimated in a single act.

The greater implications of human composting are as grand as you want to make them. In collapsing the distance between our conscious lives and certain deaths, we might live more presently. We might resume contact with the plants, animals, waters, and atmosphere we rely on to survive. We might overcome the abstractions of modernity—abstractions that have allowed us, with frightening indifference, to bring the Earth and all of its inhabitants to the brink of destruction.

Of course, NOR risks the opposite effect, too. As a matter of convenience, one might be deluded into thinking their ecological sins in life could be absolved in death. Recompose claims that each person who chooses composting over conventional burial or cremation will prevent an average of one metric ton of carbon dioxide from entering the atmosphere. According to the EPA's calculator, that is a modest carbon payback, equal to the consumption of about ten tanks of gas. On the other hand, this is preferable to adding to the debt.

To my mind, it's the perceptual shift that bears the greatest promise. If we begin to imagine ourselves as beneficial contributors to the Earth in death, rather than as agents of sickness and damage, maybe we can start to see that possibility for our lives. Put another way: we don't have to wait to die to make ourselves useful.

On the first warm day of spring, I drove to Bells Mountain. Its primary steward is Elliot Rasenick, a former music-festival organizer with a degree in religion. In 2019, he had been hoping to restore a seven-hundred-acre tract that his nonprofit had recently purchased. It was going to take an incredible amount of compost to rehabilitate a square mile of degraded land, and he wasn't sure how to get it. At the same time, Katrina Spade was trying to figure out what to do with the massive amount of soil that Recompose was about to create. The partnership seemed fated.

When I arrived at Bells Mountain, Elliot emerged from a little wooden cabin, waving. He wore rubber boots and an American flag mask. We exchanged niceties about the good weather and the elk tracks I spotted by the car. He asked what I wanted to see. I said, "Everything."

For the next few hours, we wandered around on foot. We started in the lowland conifer forest, then wound our way through a grove of ancient oaks and up a hill to a defunct rock quarry. He showed me a couple of culverts built into a stream, which he planned to have removed in the coming year, the only barriers to salmon running from the Columbia River.

Elliot was a mellow guy, and a bit of a philosopher. He told me that NOR was a fitting acronym, "a way of describing the material as something existing in a liminal state." The soil is neither spirit nor material, he said, or else it's both/and. Elliot shared my feeling that human composting's greater promise is its potential to occasion a paradigm shift in our relationship to all life. If we can understand NOR soil as sacred because of its source, maybe we will begin to perceive all biomass as sacred. That might sound like hippie stuff, but it may be what's required for our species to survive. "The climate crisis is fundamentally a soil crisis," Elliot mused. "There is a poetry in the possibility that the death of one generation can make possible the life of the next."

We climbed into a kind of all-terrain golf cart and began to ascend a narrow gravel road that hugged the sheer side of the mountain and scared the living shit out of me. I'd been reading about the inaugural dead for weeks, and pieces of their obituaries floated back to me, though I couldn't necessarily remember which story belonged to whom. One painted watercolors, another knitted "clothing for people and dogs." One spent World War II working to eliminate "social diseases" in men and kept an orange tree alive in Tallahassee "through the worst of freezes." On Earth as in the ether, their individual stories were now collective. They commingled in my mind.

We bumped along, climbing through the dense underbrush of a former clear-cut, past reed-skirted ponds you wouldn't know were man-made. We disembarked in a battered former meadow that had been used for cattle grazing. At its edge, an old burn had cleared the view all the way to Mount Adams and the flattened majesty of Mount St. Helens, still covered in snow, brilliant against the blue.

Elliot kindly turned his back so that I could cop a squat, and the hot relief of emptying my overextended bladder while taking in this view filled me with such a love of living! All that afternoon I felt alive. Walking in the shade of trees, over dead leaves and cedar fronds. And then, in that ATV thing, with rolling waves of fear giving way to adrenaline, devouring my peanut butter and jelly sandwich with animalistic fervor. Meditating on the death you want might help you imagine the life you want. But it can also help you appreciate the life you already have.

Coming back down the mountain, we stopped in another clearing, dotted with alder saplings, and cut the engine. There it suddenly was: the pile of compost that was their bodies. The first five donated yards of NOR, so unassuming and small, under that dome of sky.

We stood before the pile in silence.

Then I said, "Wow."

Elliot asked whether I wanted to place some of the soil at the base of an alder, as the Recompose staff had done during the ceremony.

At first I thought, *No.* Strangely, it seemed like an invasion of privacy to touch them. They weren't compost to me; they were people, with hobbies, and ethical convictions, and loved ones out there somewhere still grieving. They were precious.

Then it occurred to me that it was precisely this feeling that equipped me for the task.

I dug my upturned hands into the mound and lifted the soil into the sunlight. It looked and smelled exactly like the forest floor.

Ways of Knowing

JULIAN AGUON

To Hell with Drowning

FROM *The Atlantic*

I KNOW NOTHING of the night sky.

This saddens but does not surprise Larry Raigetal, a master navigator who is chewing betel nut beneath a canopy of stars. He is from Lamotrek, an outer island of Yap, in the Federated States of Micronesia. But we are meeting in a canoe house on the neighboring island of Guam, where I call home. As we speak, Raigetal is using his hands to split the horizon into a thirty-two-point star compass. He is drawing on centuries of knowledge to explain to me the art of wayfinding—a method of non-instrument navigation that has been used by his people for thousands of years to voyage between the many atolls and islands of Micronesia.

To my surprise, the compass he is conceptually grafting onto the sky is more than a map of stars as they rise and fall from east to west across the horizon. Wayfinding is a manner of organizing an elaborate body of directional information collected and committed to memory by countless navigators before him and passed down through chants to his grandfather, to his father, to him. It's a living repository of spectacularly specific details about sea swells, wind currents, reefs, shoals, and other seamarks—including living ones. A pod of pilot whales. A shark with special markings. A seabird.

As a Pacific Islander, I knew that the canoe house has long been a place of learning, and I'd come to ask Raigetal about whether wayfinding had been compromised by climate change. As a human-rights lawyer working at the intersection of Indigenous rights and environmental justice, I'd also come because I believe that the peoples of the Pacific have important intellectual contributions to make to the global climate-justice movement. We have insights born not only of living in close harmony with the Earth but also of having survived so much already—the ravages of extractive industry, the experiments of nuclear powers. We have information vital to the project of recovering the planet's life-support systems.

Finally, I'd come because my personal and professional reserves were depleted. Like so many others working in the climate space, I'd been feeling overwhelmed since August, when the Intergovernmental Panel on Climate Change (IPCC) released part of its sixth assessment report. The conclusions were bleak. Reading the report felt like being buried alive by an avalanche of facts—the facts of sea level rise and progressively severe storms, among others— and I was looking to claw my way out.

As the darkness deepened around me and Raigetal, I realized two things. First, the climate-justice movement must listen more carefully to those most vulnerable to the ravages of climate change, such as Oceania's frontline communities. Second, we who are waist-deep in that movement need more than facts to win. We need stories. And not just stories about the stakes, which we know are high, but stories about the places we call home. Stories about our own small corners of the Earth as we know them. As we love them.

In my corner, Micronesia, the facts are frightening. We are seeing a rate of sea level rise two to three times the global average. Some scientists theorize that most of our low-lying coral-atoll nations may become uninhabitable as early as 2030. Faced with the prospect of climate-induced relocation, some leaders have contemplated buying land in other countries in anticipation of having to move some or all of their people.

One leader has already sealed a deal. In 2014, the then-president of Kiribati, Anote Tong, entered into a purchase agreement with the Anglican Church for more than five thousand acres in Fiji, paying nearly $9 million for them. (Kiribati has since begun using the land for farming.) Though the deal was seen as visionary by some, to others it marked a kind of death. After all, at what point does an agreement that envisions the relocation of an entire human population—now some 121,000 people—become more eulogy than contract?

In Fiji, the government keeps its own kind of death list—an official record of all the villages that may have to be relocated because of sea level rise. Using internal climate-vulnerability assessments, the

Fijian government has determined which of its coastal villages are most susceptible to coastal erosion, flooding, and saltwater intrusion. As of 2017, forty-two villages were on the list. If and when they are forced to move, they won't be the first: in 2014, Vunidogoloa formally relocated to higher ground, some two kilometers inland.

When I spoke with Sailosi Ramatu, that village's headman, in July, he told me the move was hardest on the elders. In the months leading up to the relocation, they held prayer circles. They fasted. They readied themselves for the rupture of having to abandon their ancestral lands. In Fiji (as in many of our islands), the people are tethered to the earth, as enshrined in the concept of *vanua*, a word that means "the land" and "the people" at once. Vunidogoloans live and love and die on their lands, most of which they do not even own, at least not as individuals. Rather, theirs is a system of communal land ownership. They tend to their gardens. They bury their dead. They even bury their umbilical cords. So, it was no surprise when, as about thirty families set out for the new site, some of the older women wailed as they walked.

Perhaps that's a sound the sea makes when it rises: old women wailing.

Not everyone made the journey; the dead remain interred in a cemetery at the old site. According to Ramatu, one of the biggest struggles his people faced was leaving their buried loved ones behind. Some worry they'll be cursed for abandoning their deceased relatives. Others walk around with holes in their heart. Like the old man who visits the cemetery nearly every day to sit by the grave of his dead wife. I'm not sure which flowers he brings her, if any. But I imagine they're beautiful.

Perhaps the story of climate change is a story of flowers.

These are the facts in the Republic of the Marshall Islands (RMI), the country where the U.S. military houses its Ronald Reagan Ballistic Missile Defense Test Site: there, a crucial study on sea level rise found that coral-atoll nations may not be able to sustain a human population past the present decade. This conclusion was met with trepidation by the Marshallese people I spoke with, who hear the ticking of the climate clock louder than most.

The 2018 study, led by the United States Geological Survey and commissioned by the Pentagon, focused exclusively on an island in the Kwajalein Atoll that supports some 1,250 American military personnel, contractors, and civilians living there and on nearby islands. For the most part, the United States otherwise ignores this region. Wake Island, where an additional study on sea level rise is now being done, is proof of that fact.

Wake, an island with no permanent inhabitants and which the United States considers an unincorporated territory, is run by the air force under authority of a caretaker permit issued by the Interior Department. For its part, the RMI not only has a competing vision of what caretaking looks like; it also has a competing claim to Wake. In April 2016, the RMI formally claimed Wake Island when it filed its maritime coordinates with the United Nations secretary-general.

The truth is that neither government is entirely correct. The strongest claim is that of the Marshallese people themselves, who say the island is theirs by way of history, culture, and birthright, and who long to be able to take proper care of it. They also say that Wake is not the island's true name.

Its true name is Enen-Kio. The island of the orange flower.

Famous in lore for the beauty of these flowers, Enen-Kio is also known for its rare assemblage of nesting seabirds—frigates and albatross, among others. Legend has it that local warriors, seeking to prove their worthiness, would journey to the island in search of the wing bones of one such seabird. Fourteen years ago, on another starry night, a high chief explained to me that the retrieved bones were used as chisels in traditional tattoo ceremonies.

I did not grasp the significance of the strip of orange splayed across the RMI flag until much later. Former president Hilda Heine would tell her poet daughter, Kathy, who would tell me: For the Marshallese, orange is the color of bravery.

On my island, climate change is a story of storms. Guam—the largest and southernmost of the Mariana Islands and an unincorporated territory of the United States—lies within one of the most active regions for tropical cyclones in the world. The typhoons that have historically battered the island are so strong, they're often called "super typhoons."

Everyone here remembers their first. Mine was Omar, in August 1992. We were unprepared—my mother, brother, sister, and me. This was in part because my father, who typically did the preparatory work of putting up shutters and removing debris from around the house, had recently died. I remember the four of us huddled behind a cream-colored mattress. I remember tracing its embroidered flowers with my finger.

I remember everything, really. Trees and telephone poles cracked in half. The roof of our neighbor's house went flying, as did his canopy and one of his cars. I remember glass everywhere, as several windows and a sliding door shattered. I remember the sound of the wind as it blew under the bottom of my bedroom door. Like an old man sucking his teeth.

Pamela is the one my mom remembers. May 1976. One of the most intense storms to strike Guam last century, Pamela generated eight-meter waves and ravaged the beaches on both the northern and eastern sides of the island. She sank ten ships in the local harbor. She did an estimated $500 million worth of damage. But none of this is what my mom remembers. What she remembers, what she will never forget, is a single white toilet. American Standard. The one thing left of her house when Pamela was over.

Then there was Paka. December 1997. The wind and rain beat down on us for twelve hours. The barometric pressure was so low that it was believed to have induced labor in nine pregnant women. Paka, like Russ in 1990 and Yuri in 1991, unearthed untold numbers of dead bodies when it slammed into the southern cemeteries of Yona and Inarajan. Corpses spilled out of their coffins. Coffins bobbed like buoys in the bay.

Several families spent weeks combing the beaches in search of their loved ones. Some were never found. My aunt, who worked for one of the cemeteries, said that one family was able to identify their father's body only because of a cherished baseball cap, which they had buried him in and which had stuck to his skull by way of a mess of seaweed. Suffice it to say, when the IPCC dropped its latest report, confirming that tropical cyclones are just going to get stronger, my corner of the world shuddered.

After all, although 1.5°C of warming will make these storms even more severe, that same severity will increase dramatically with 2°C let alone 3°C. I can't begin to imagine what any of this will mean on the ground—and not just for Guam or the Northern Mariana Islands, but for Vanuatu, Fiji, and the Solomon Islands, whose communities already seem to be lurching from one Category 5 cyclone to another.

*

Throughout Oceania, the story of climate change is also a story of ingenuity. In the Carteret Islands—off the coast of Bougainville, in Papua New Guinea—the women are taking matters into their own hands. Frustrated by how slowly the Papua New Guinean government was implementing its relocation plans, they sought to mobilize the community around the issue of relocation. They formed an organization and named it Tulele Peisa, which means "sailing the waves on our own" in the local Halia language.

To date, Tulele Peisa has organized several community consultations as well as visiting missions between the Carteret Islanders and potential host communities in nearby Bougainville. According to Ursula Rakova, the group's leader, Tulele Peisa has also secured several tracts of arable land on which it is now growing gardens of taro and cassava. She told me the group has planted more than thirty thousand cocoa trees and even established a cocoa-bean refinery. All told, Tulele Peisa has developed an eighteen-point relocation plan for its community. Rakova said other coastal communities have followed suit and are currently formulating their own relocation plans.

So, it would seem that the Pacific Climate Warriors were right. The youth-led group fighting climate change across Oceania, as part of the global 350.org network, famously declared: We're not drowning. We're fighting.

And we are.

The Marshall Islands has spearheaded the Climate Vulnerable Forum, a group of forty-eight countries that works to amplify voices that have long been marginalized in the climate realm. Fiji, too, has taken a leadership role, presiding over the twenty-third Conference of the Parties to the United Nations Framework Convention on Climate Change and spearheading the so-called *talanoa* dialogues—sessions that use storytelling to foster more empathetic decision-making.

One could argue that Fiji is also leading the way on the complex issue of climate-induced relocation. Consider the list of forty-two villages slated for possible relocation. However heartrending, the very existence of the list is a testament to Fiji's efforts. Conversations about relocation are enormously difficult to have, but that country is having them.

Tuvalu, Tokelau, Kiribati, and the Marshall Islands have joined forces with the Maldives to form a coalition of coral-atoll nations to advocate for the financial resources necessary to adapt to climate change. To date, what money most of them have been able to secure has been limited to funding the first-generation stuff of seawalls and early-warning systems—nowhere near the level they will need to actually adapt, let alone adapt in place. But they press on, planting mangroves and plugging away at their national plans.

A group of law students from the University of the South Pacific led a different charge in Vanuatu: they advocated that that climate-vulnerable country take the lead in pursuing an advisory opinion on climate change from the International Court of Justice. As these students see it, the lack of clarity around state duties is impairing the collective efforts of the international community to respond effectively to the climate crisis. In September, the students succeeded in their initial goal, and Vanuatu announced that it would spearhead the initiative. (I should note here that I am leading the global team assisting Vanuatu in this effort.)

In the Marianas, many Indigenous Chamorros and Carolinians of Guam and the Commonwealth of the Northern Mariana Islands are fighting the destruction of our lands and seas by the single largest institutional producer of greenhouse gases in the world: the U.S. military. On land, these activists are opposing the construction of a massive firing range, which they say will destroy a limestone forest and imperil a whole host of nonhuman life. At sea, they're challenging the Defense Department's attempt to militarize a section of the ocean almost the size of India.

All this to say, if my corner of the Earth had an anthem, it'd be this: to hell with drowning.

That anthem was never more clearly on display than during the twelfth Festival of Pacific Arts, held in the summer of 2016, when the *Lucky Star* arrived at the local harbor. *Lucky Star* was one of three canoes sailed to Guam from Lamotrek by the wayfinder Raigetal and his crew of apprentices. What made this particular canoe so special was the fact that its sail was traditional, meaning it was woven from pandanus leaves by the women of Lamotrek.

There, in the weeks leading up to the voyage, it was discovered that the knowledge of how to weave such a sail was very nearly lost. Literally one woman still knew how to do it.

Her name was Maria Labushoilam, a ninety-year-old master weaver, and she was dying.

Maria would spend the last two weeks of her life teaching fifteen women how to make that sail. From her deathbed, she taught them how to harvest, dry, and split the leaves, and then how to weave them. After she died, the women completed the sail without her. The community raised it together.

The Lamotrekese are emblematic of the predicament all Oceanic peoples are facing today: we come from cultural traditions rich in beauty and resilience—the same traditions that have enabled us to thrive in our ancestral spaces for thousands of years—but that is simply not enough to ensure our continued survival. The part simply cannot save the whole. The answer to the question of climate change must come from everyone, or it will come from no one.

You could say Maria performed something of a miracle in her final days. With nothing more than pandanus leaves and love, she opened a window to a world—a future in which good people refuse to simply lie down and die, a future rooted in respect for possibility, a future with room for us all.

May we have the courage to climb through it.

CLAUDIA GEIB

To Speak of the Sea in Irish

FROM Hakai Magazine

SITTING AMID THE BRIC-A-BRAC of generations of seafarers before him, fisherman and museum curator John Bhaba Jeaic Ó Confhaola of Galway, Ireland, tried to describe a word to interviewer Manchán Magan. The word, in the Irish language, was for a three-bladed knife on a long pole, used by generations of Galway fishermen to harvest kelp. Ó Confhaola dredged it from his memory: a *scian coirlí*.

"I don't think I've said that word out loud for fifty years," he told Magan.

It was a sentiment that Magan would hear again and again along Ireland's west coast. This is a place shaped by proximity to the ocean: nothing stands between the sea and the country's craggy, cliff-lined shores for roughly three thousand kilometers, leaving it open to the raw breath of the North Atlantic. Many cities and towns here have roots as fishing villages and ports, and for generations, to speak Irish in them was to speak of the sea.

A sarcastic person might be described as *tá sé mar a bheadh scadán i dtóin an bharraille* (like a salted herring from the bottom of a barrel). To humble a braggart was *an ghaoth a bhaint as seolta duine* (to take the wind out of their sails). Each community developed its own vocabulary: words for every sort of wave, every tide, and every shift in weather; for the sea's sounds, its plants, and its creatures; and for the tools and tricks a mariner used to make a living on the ocean's surface.

Yet this unique vocabulary is slowly disappearing. Early last year, Magan—a writer, documentary filmmaker, and connoisseur of the

Irish language—began collecting coastal words from towns along the west coast, in an effort to preserve them.

"I was hearing these words from fishermen with these concepts, these connections with the other world, that were really profound—and now they're no longer being said," Magan explains.

Supported by funding linked to Galway's designation as a European Capital of Culture for 2020, Magan spent February and part of September recording stories and sayings in Ireland's Atlantic communities. The recordings make up the Foclóir Farraige, or Sea Dictionary: an online database of recordings and definitions sorted by their regional origin. Magan also recently published a selection of words in an illustrated book.

Some of the words of the Foclóir Farraige are functional. They describe tools that mariners once needed, like the *scian coirlí*, or the *strapa ballachaí*, a Galway word for rope strung through the mouths of up to thirty wrasse and formed into a loop, to carry the catch home.

Yet the words are often much more than utilitarian. They carry a sense of poetry, and a perspective on nature. There is the town of Donegal's *mada doininne*, a particular type of dark cloud lining the horizon that foretells bad weather. The word, literally translated, means "hounds of the storm." Or *bláth bán ar gharraí an iascaire*, a description of choppy sea from the county of Galway that means "white flowers on the fisherman's garden."

Magan gave his dictionary an alternative name: Sea Tamagotchi, for a game popular in the late 1990s that challenged users to keep digital pets alive. It's a reference to Magan's hope that the dictionary might inspire people to adopt and nurture some of these largely forgotten words.

Linguists and historians consider Irish to have been in decline since 1603, when English colonialists defeated Ireland's last chiefs. Yet the events that truly reshaped Ireland's coastal vernacular didn't take place until 1973, when the country joined the ranks of the European Union.

At the time, the country was struggling. Poverty and unemployment ran rampant. Use of Irish had already dwindled to a few small pockets, primarily along the west coast, known as the Gaeltacht regions. After the Great Famine of the mid-1800s, the language had become associated with poverty, trauma, and backwardness.

English opened doors to better work, including jobs abroad: "You needed English to get out," Magan says.

With its entry into the EU, Ireland agreed to a shared fisheries policy, giving EU member states equal access to Irish waters. The resulting quotas were largely responsible for driving much of the Irish fishing industry out of business. And as the fishing industry waned, so did its terminology. "What happens is, if you don't maintain the economy of fishing, or farming, or peat cutting, then the words connected to them die with it," Magan says.

Old fishers and mariners became the only remaining keepers of a vocabulary with limited use, and one which is no longer passed down. To some Irish speakers and observers, this is a profound loss.

"Languages are your window on the world," says Brenda Ní Shúilleabháin, a lecturer and Ph.D. candidate in history at Ireland's University of Limerick, who has documented Irish life through books and film. While some Irish sensibility might translate into English, the loss of many words and the common use of the Irish language itself means people "now view [the world] through a somewhat different medium," she says.

Today, however, the tide may be changing. The twenty-first century has seen a resurgence of interest in the Irish language. In Ireland and beyond, Magan recognizes a hunger for cultural connection that the Foclóir Farraige could fill. It's a thesis he's tested before: in past exhibitions of his work, Magan gifted visitors with other disappearing Irish words for protection. Over the years, he learned that recipients carried their words in their wallets, bestowed them on pets and beers, or in a couple of cases, had them tattooed on their bodies.

"All around the world, we feel uncertainty and want to connect with what roots us to heritage, to the past, something that has meaning," he says.

The Irish language reflects a deep relationship between humans and the natural world, a sensibility shared with many Indigenous languages—from those belonging to First Nations of British Columbia to the Ainu of Japan. Irish largely does not demarcate between the human world and nature, nor between this world and the next.

A coastal Irish speaker, walking the beach at night, might have equally expected to hear *stranach* (the murmuring of water rushing from shore), or the whisper of *caibleadh* (distant spirit voices drifting in over the waves). They knew the *ceist an taibhse* (the question for the ghost)—a riddle used to determine if someone they met along the way was human or supernatural. Many words describe ways of predicting the weather, or fishing fortunes, by paying attention to birds or wind direction; to the sea's sounds; or to the colors in a fire.

To Donnchadh Ó Baoill, a Foclóir Farraige contributor, this everyday magic has its place in modern life.

"We live in a very fast world, where we often give things a very quick glance," says Ó Baoill, a former language and culture officer for the Gaeltacht region of Donegal. The Irish language helps speakers see details they'd otherwise miss, he says. "That detail enriches our lives. And the landscape becomes alive."

Few people, including Magan, expect that the words of the Foclóir Farraige will return to everyday use. As Ní Shúilleabháin puts it: "I'm a realist; every language changes." Yet she also sees the urgency in Magan's work, as dominant languages subsume smaller tongues around the world.

"In an increasingly homogenous world . . . I think it's very important to maintain cultural intimacies," she says.

Ó Baoill and Magan both point out that preserving Ireland's traditional coastal vocabulary is especially important in the face of climate change and biodiversity loss. Take a word like *borráite*, from Carraroe village, which describes a rocky offshore reef found in the area. Kelp once grew on these reefs in abundance, tangling with other seaweed species and providing refuge for fish. Due to climate change and overfishing, however, Magan says that a borráite today would host neither kelp nor many fish. "Contained within that word is the entire ecosystem that was in that area," Magan says.

Words like this, he hopes, can both remind us of what we have lost and reconnect us to what we might still preserve. One doesn't need to speak the language to understand such a message.

MIKKI K. HARRIS

A Tight-Knit Island Nation Hopes to Rebuild While Preserving "The Barbudan Way"

FROM *National Geographic*

I'VE HAD A RELATIONSHIP with water since before I was born.

As a young child at a barrier island off the coast of southern New Jersey, my father threw me into the cold open waters of the Atlantic, an exercise for me to find my bearings inside the waves. Watching intently, cautiously, he repeated the words that Papa, my grandfather, spoke to him: *The ocean is like anything in life. Learn its rhythm and you'll not only flow with it; you won't have to work so hard to enjoy it.*

My father is a sixth-generation descendant of Barbudan people. I am a seventh.

The skill of navigation, the art of catching fish, understanding the meaning of the temperature and color of the water, and so much more has been passed down to me and others through our people. And being connected to this independent, self-governing Black community, having been immersed in this culture, has been central to my identity as an African of the Americas.

It's not surprising if you've not heard of Barbuda: this small, sixty-two-square-mile island in the eastern Caribbean is the less known half of a twin island state that, along with Antigua, gained independence from England in 1981.

Barbudans have kept mostly to themselves, concerned less with growing tourism than with working to preserve their land and maintain a tight-knit, ecologically sound community of about 1,500 people.

My grandparents were born and raised in the very same village as my great-great-great-great-great-grandparents. Like my father before me, I learned about my connection to the Earth and her seas at the Codrington Lagoon, the wetlands that serve as the center of community life and entrepôt for our fisheries.

Barbudans are descended from West Africa and Africans of the British Isles, with many—like my family—tracing lineage back nine generations. We are fishers, navigators, farmers, and artisans with a way of life that has remained largely unchanged for more than three centuries.

Long before the British set foot on Barbudan shores, the land was known by the Caribs, Siboney, and Arawak Nations as Wa'Omoni. The presence of First Nations peoples can still be found in Barbuda today through genetics and a drawing on the interior wall of "Indian Cave" on the northeast end of the island. The British Crown claimed the island in 1685, after which Scottish brothers Christopher and John Codrington were granted an initial lease to the island by King Charles of Great Britain. This lease would later be extended by Queen Anne once she was private owner of the island. Christopher managed his sister's plantation, Betty's Hope, sixty-three kilometers away on Antigua, and had a plantation in Barbados, but was not able to create a similar plantation on Barbuda because the terrain was not conducive.

The Codringtons looked to the British Isles to import skilled laborers and indentured servants. Those who arrived on Barbuda via the main port on Antigua were skilled as hoopers, coopers, shipwrights, metalworkers, and sailmakers. A number of freed Blacks who lived in Europe were drawn to the offer because in addition to wages commensurate to their level of skill, a plot of land for a domicile, a plot of land for provisions—known as "grung" even today—and supplementary rations were also offered. These laborers—plus a handful of indentured whites—were the first Barbudans. The white indentured servants departed once their debts were paid, but the Black laborers from the British Isles and later West Africa remained, and their ancestry is connected to Barbudans today.

Barbudans are an extension of the land. It's difficult to fully express through the written word, or even in photographs. The land in many respects *is* us. In a letter from June 1, 1834, Christopher Codrington described Barbudans as "one united family so attached to Barbuda that force alone or extreme drought . . . can alone take them from that island!"

Threats to Communal Well-Being

Throughout the world, communal well-being has long been a feature of free Black communities, especially those that grew from resistance to the international slave trade. Jessica Gordon Nembhard, author of *Collective Courage: A History of African American Cooperative Economic Thought and Practice*, notes that every human population in every era of human history has used and uses mutual aid, economic solidarity, and cooperation to survive and thrive. She found that African Americans, like other subaltern peoples, used multiple forms of economic cooperation and collective ownership to resist enslavement and free themselves, to feed their families, to maintain land ownership, to secure affordable housing and nonpredatory lending, to create decent jobs, and to keep resources recirculating in their communities.

Yet from the Gullah Geechee of the coastal American South to the Maroons in the mountains of Jamaica to the Garifuna of Latin and Central America, remote Black communities with their own distinct identities, language, and culture have been under relentless siege by the threat of commercial development. Unless there is international attention to quickly unify and support the rights of the Barbudan people, this independent, culturally distinct community will be destroyed.

Born in 1890, my great-grandfather James "Boatie" Harris, a Barbudan fisher and ship caulker, could navigate the dark using only the lights from the village and the reflections from Antigua to determine the course for a boat. The fish that Boatie caught would feed his family and those in the village. Barbudan fishers decide how often they fish the lagoon and sea, know what type of fish to remove from the water, what time of year, and what size they have to be—all decisions that secure the sustainability of the fishing stock.

But fishing "the Barbudan way" is threatened by the potential shift in accordance with profitability, absent the best interests of the people. The link among people, land, resources, lifestyle, and culture is based on living within the resources of the community.

To live on the island, Barbudans call on traditional knowledge to use and to protect the resources they have access to. They use their skills as fishers, hunters, and farmers for food security. There is a freedom for kids to chase donkeys and splash in the lagoon, and a tradition for families to camp and cook on the coastal shore during summer months. Once one person starts a fire to cook, everyone has fire to cook. This way of life has kept not only families intact but the ecology secure.

According to Cynthia Hewitt, sociologist in the political-economy tradition of world-systems theory, and director of International Comparative Labor Studies at Morehouse College: "There is an assumption that development means greater material good, and with that, greater money. But with that development is really a shift in the community well-being."

Shift in Centuries-Old Methods

Hewitt speaks of a shift, and for Barbuda, the introduction of development for and by non-Barbudans moves the people from co-operative owners to dependent workers. What's more, as land and other natural resources are developed to lure tourists, the makeup of Barbudan society changes, and a delicate ecology is disrupted. These shifts can erase centuries-old methods of living with the land in the name of development.

For more than two hundred years, Barbudans operated with a system of communal land ownership. In 1976, the Barbuda Local Government Act formalized the administration of communal ownership of the island with the creation of the locally elected Barbuda Council. In 2007, the then central government administration in Antigua passed the Barbuda Land Act, which states that every adult Barbudan holds the land in common, with the Barbuda Council serving as administrators. It also explicitly specifies that when it comes to leasing land for major developments, consent must be obtained from a majority of the Barbudan people through an in-person vote.

A decade after the landmark legislation, Hurricane Irma tore across the island with 185-mile-per-hour winds, eroding shores, and bringing level-1, -2, and -3 damage to what the Caribbean Disaster Emergency Management Agency reported was 99 percent of the infrastructure. Then, when yet another massive storm was predicted to make landfall days later, the Antigua-Barbuda central government (akin to a federal oversight body) ordered the island evacuated. When Barbudans were finally allowed to return, nearly everything had been destroyed.

A core group of Barbudans vowed to rebuild, wanting to preserve their cultural space. But it was hard to know where to start. Islanders were initially buoyed by an Antigua-Barbuda government that arranged for the swift delivery of construction equipment; but instead of repairing homes and schools and basic infrastructure (like the hospital and post office), we watched in shock as bulldozers started clearing even more trees and land to make space for a new, international airport.

While the chairman of the locally elected Barbuda Council signed off on this deal, many say he failed to follow the mandated three-part process to obtain approval from the Barbudan people. Then, six weeks after the storm, while Barbudans were still picking through the wreckage of their homes by hand, an even larger, 650-acre development plan was approved by the Antigua-Barbuda government without involving the local council, or the Barbudan people, in what many view as a land grab. Prior to the 2017 storm, the government had approved a ninety-nine-year lease to the Peace Love and Happiness (PLH) partnership, a company co-owned by billionaire businessman John Paul DeJoria, founder of Patrón tequila, for the construction of a resort and commercial village on 425 acres.

As if these actions weren't disturbing enough, the following spring, in 2018, the central government did something that brought attention to human rights advocates worldwide: they voted to repeal the Barbuda Land Act, stripping Barbudans of our communal land rights, a central feature of our people for centuries.

Juliana Nnoko-Mewanu, a senior researcher on women and land for Human Rights Watch, authored a 2018 article on Barbuda, citing how research consistently shows that "taking away land used by communities—without due process and without adequate compensation and rehabilitation—results in serious risks to people's rights to food, water, housing, health, and education.

"Barbuda may be small," Nnoko-Mewanu concluded, "but the rights of its people are as important as anyone else's."

While some Barbudans undertook the physically grueling work of clearing broken concrete and debris from damaged homes and government buildings, others started shouldering the emotionally and financially burdensome process of fighting in court. While they have received some assistance from international groups concerned about irreparable environmental destruction, expenses outweigh the community's ability to pay, and they are relying on grassroots fundraising—all while many are still displaced and living in tents or staying with relatives on Antigua.

Hoping and Praying

Renetta "Brownie" Nedd, ninety-five, has a three-bedroom, Caribbean Sea–blue home with wooden hurricane shutters that cover the four front windows and door that face the large water well known in the village as "Park Well."

"Irma wasn't easy. It wasn't easy. People crying mercy. Thank God He did have mercy and He is still having mercy because if He did not have mercy we would not be here today," Nedd said when retelling her experience after the storm. She took shelter with one of her sons while Irma battered the island. The next morning, she walked to check on her home, and did not recognize the house that she and her husband James "Hercules" Nedd updated from wood to concrete in the 1960s. She spoke of the community, and the way that homes were built and fixed on the island. "When people building a house, they have young fellas they get together. I help you, you help me, and that's how they used to work. So, I think they still have a little of that in them."

Fifteen weeks after the storm, on Christmas morning, the village united to clear and prepare Nedd's home for a new roof. The effort was led by Sean Charles and Mike Harris, two Barbudan retired U.S. veterans, who returned home to unite Barbudans and to prepare buildings for repairs.

As the two-year anniversary of Irma approached, Nedd said, "I have life. Where there is life, there is hope. And I'm hoping if God's will, I may be able to get back in there. I'm hoping. I'm praying."

She left the island in July 2019 to visit a daughter in Saint Thomas. By the time she returned in December 2020, the roof was completed. Community volunteers also installed a bathroom and flooring for the house. She has since moved back in. Electricity is back on, and kitchen repairs are underway.

Battle Against Development

Currently at least five separate lawsuits address the questions of Barbudan sovereignty, deforestation, and damage to wetlands, protected since 2005 under the RAMSAR Convention. The lawsuits question the legitimacy of recent changes to laws in order to benefit developers such as DeJoria's PLH, Robert DeNiro's Paradise Found, and Discovery Land Company, an Arizona-based developer. Yet while the court battles wind on, PLH has added additional development partners and expanded plans to include two mammoth private residences on about 114 acres directly inside the Codrington Lagoon National Park, the heart of Barbudan life and a hub for industry and recreation. Damage to the wetland ecosystem also continues.

In multiple public statements, PLH claims its projects are as much about "giving back to the environment" as they are about opening the island to commerce.

Yet conservation scientists unaffiliated with the island have vehemently objected to the idea that destroying protected wetlands to build a private golf course and waterfront homes is ecologically sound.

"The proposed development would deal a mortal blow to Barbuda's fisheries," the president of the Global Coral Reef Alliance, Thomas J. Goreau, a marine biologist and biochemist, wrote to fellow scientists in May 2020. "The impacts on Barbuda would be . . . severe [and] destroy the most important mangroves, seagrass, coral reefs, and fisheries of an entire island of fisherfolk."

Money Over Matter

Antigua and Barbuda prime minister Gaston Browne has repeatedly said that he wants Barbuda to "make a net positive contribution to the treasury." Instead of acknowledging the concern for the continuance of Barbudan identity and traditions, he speaks of members of the current Barbuda Council as preventing growth and has bluntly acknowledged his willingness to "go to Parliament to change the law to facilitate the [PLH] project."

In 2018, he oversaw the repeal of the Barbuda Land Act of 2007 through the Crown Lands Regulation Amendment Bill of 2018, which allows for the private ownership of land by foreigners, and the enactment of The Barbuda Amendment Bill of 2018, which gives the central government the authority to approve major development on Barbuda without consultation and consent of the Barbudan people. The bill also removes the ancestral identity of Barbudans, referring to the "inhabitants" as tenants. Historically, Barbudans are defined as direct descendants of the original Barbudans who came from West Africa and the British Isles, and in order to serve on the Barbuda Council, or hold the land in common, a person had to be Barbudan. No one could purchase Barbudan identity, nor ancestral rights to the land, but the changes to the law in 2018 roll out a red carpet for privatization that removes agency from the people.

To be sure, when it comes to paying for storm repairs and upgrading the island's infrastructure there are no easy answers. There are Barbudans who have gone to work for developers, viewing the immediate benefits of employment—the means to put food on the table—more urgently than the ideological aim of ensuring our ethnic group's survival.

But allowing the continuation of a development plan that strips the voice of the Barbudan people all but ensures a historic, culturally unique Black ethnic group will be erased.

A Legacy Worth Fighting For and Preserving

Teckla Negga Melchoir, a Barbudan anthropologist and journalist, and founder of Global Forward Thinking.org, "estimates [that] one hundred million First Nations persons have been exterminated in the Americas since the Europeans began 'discovering' the rest of the world. Barbudans are thought of as 'just fifteen hundred people on island.' Since the forced evacuation after Hurricane Irma, the local population has been decreased by at least three hundred. Many of whom were denied a right to return home as they were denied the right to make a living because of the illegal and draconian 'laws' passed by the current administration in Antigua, laws that facilitate foreigners who have no connection to the land, no blood in the soil, and no concern for the environment, culture, or people. These foreigners like those centuries ago happened upon Paradise and decided that it was up for sale and that it could be improved."

She continues, "You ask me about development? Development for whom? Every Barbudan born in Barbuda, usually at home, has their umbilical cord buried in the yard of their house. That navel string feeds the land, roots us to the land and the land feeds us. It doesn't matter where in the world we Barbudans go, and there are hundreds upon hundreds of us, we are emotionally, psychologically, and intellectually tethered to Barbuda."

Prominent Black thinkers often lament how gentrification causes the loss of historic Black spaces and Indigenous cultures, the destruction of Black property, the lack of generational wealth.

Barbuda's soil survived the impact of Hurricane Irma because of the strength of the coral reefs and wetlands. Her people likewise survived the storm, but the powerful forces of economic "progress" may wipe out a homeland, a culture—an identity—once and for all.

The fate of Barbuda should rest in the hands of the Barbudan people, but now the global community is needed to recognize the distinct island culture and a people whose human rights are being violated. Without wide support in this era that the United Nations General Assembly has called "The International Decade for People of African Descent," Barbudans, who offer important direction to a collective future are in danger of becoming just an interesting memory if we don't recognize and respond to the urgency of now.

Barbuda is not just land. It is a Black diasporic identity. It is 1,500 lives that represent a freedom tradition, a legacy worth fighting for and preserving.

'CÚAGILÁKV (JESS HÁUSTI)

Thriving Together: Salmon, Berries, and People

FROM Hakai Magazine

WHEN I WAS SMALL, my ǧáǧṃp (grandfather) would set about the serious business of food gathering with my cousins and me in the late spring. Everyone in the family had a role in our food harvests and backyard cannery, and the children's role came early in the salmon season. As children, we believed the whole success of the harvest, not only of berries but also of the salmon that soon followed, depended on our performance. Our ǧáǧṃp would furnish us with buckets, hammer nails into the ends of long, split cedar sticks, and gravely send us off on a mission to find ǧúláli (salmonberries). We'd seek out the best bushes around the village where we live, searching for the raspberry-like berries that thrive, as we do, in the bright and salty transition spaces between Pacific Ocean spray and coastal temperate rainforest.

We'd come home when we ran out of daylight and pile our buckets on the kitchen table. Our ǧáǧṃ (grandmother) would pour the berries into a bowl—a mixing bowl in a poor berry year, or the huge ceramic bowl she used to knead the family's bread in a good one—and we would dutifully recount where the best berries had been, how big they were, and how ripe and juicy. We would speculate on how the berries compared to the previous year's and regale our ǧáǧṃp with stories about getting lost in the thickets and fending off rez dogs with the long, hooked sticks meant for pulling down high branches.

My favorite moment came in the years when my ǧáǧṃp would

nod to himself and make the official pronouncement: "It's going to be a good year for salmon." In that moment, we felt like little harbingers of hope.

For most Haíɫzaqv (Heiltsuk) children, the relationship between salmon and salmonberries is the first indicator—a sign from the natural world—we are taught. A good crop of salmonberries, we are told, corresponds to a good salmon run and luck in the harvest, and a poor crop is an early signal that we should turn to other species for our winter stores. We learn about this nourishing interrelationship early in our lives and it goes on to pattern our worldview.

Salmonberries glisten like small bursts of orange and red fish roe, nestled in the greenery beside magenta flowers and the hard, green clusters of berries still to ripen. On these shrubs, at the height of the season, you can see a whole life cycle painted across the riverbank in jewel tones. The salmonberry, from the same genus as raspberries and blackberries, has fruits that are composed of a chaotic heap of juicy drupelets that set a table to nourish a whole host of human and nonhuman kin: songbirds, small mammals, and black and grizzly bears. And the delicate fragrance and flavor are as satisfying as the dull thud of berries hitting the bottom of my bucket.

I treasure so many gifts from the salmonberries that help me through every season of the year, and my life: the fresh leaves that helped me through childbirth, the new shoots in the spring that I gently peel before eating them like licorice strings, the deep blush of blossoms that give me hope in the dark of early spring. And of course, the berries that talk to me, lovingly, of salmon as I fill buckets and bowls to make jelly for my precious ǧáǧm̓. Salmonberries are my definition of comfort food.

Salmonberry ecology aligns beautifully with the spaces that my ancestors loved to be in. The plant thrives in cool, moist coastal forests and along the lush banks of streams and rivers that pulse like deep, green arteries through our homelands. If you find a place where salmonberries, salmon, and clear, fresh water overlap, you will also find culturally modified trees—usually western red cedar, carrying the marks from planks or cedar bark strips harvested without harming the living tree—and other love notes from our ancestors left on the land for us. As Robin Wall Kimmerer writes in her book *Braiding Sweetgrass,* "all flourishing is mutual." And here in Haíłzaqv territory, all these elements—people, place, salmon, and salmonberries—can be found surviving or thriving only through our mutual care: we observe the bloom and abundance of flowers and berries as we await the coming salmon, then we Haíłzaqv, and other mammals of the territory, fertilize those salmonberry bushes with salmon remains so that they will bloom and bear fruit again in a cycle much deeper than any one season.

In my mind, salmonberries have always embodied community: their flowers nurture pollinators and their berries feed creatures of every size, winged and limbed. Salmon, laid at their feet, attract teeming insects to nestle into the soil and among the fallen leaves in the undergrowth. The space the plants hold invites you, as poet and essayist Wendell Berry writes, to "put your ear / close, and hear the faint chattering / of the songs that are to come." In their ecology, their poetry, and their lessons about reciprocity, wild salmonberry thickets, and the salmonberry gardens we actively tend, are home to diversity and abundance; we are fortunate to have so many pathways to understand their gentle might.

Increasingly, I meet Western scientists who recognize the power of Indigenous knowledge systems: they factor this knowledge into their study designs, include Indigenous knowledge keepers on their field teams and academic papers, and have the humbleness to recognize the biases they bring to their work. This is a welcome shift from how generations of Western scientists brought oppressive and extractive research practices to our territories, ignoring the wisdom of Indigenous stewards. Indigenous science, in contrast, is kincentric and relational: it is strengthened by the interdependence of human and nonhuman kin that together form wise systems. It is of deep importance in these geographies that Western science begins to reflect the patterns of interrelationships that give structure to both the culture and ecology of this place.

Western science is a curious little sister on this coast, mapping ideas and observations in spaces where Indigenous science has been foundational to kinship-building and ecological balance for millennia. As Indigenous stewards and scientists, we have much we can teach this little sister. Her curiosity, her fresh eyes sometimes show us things in a new light. And often, Western science affirms the stories and knowledge that Indigenous peoples, like the Haíɫzaqv people, have meticulously tended as living bodies of collective learning since time before memory. Taken together, we can sometimes map bigger patterns than either sibling could see alone.

A recent paper published in *Ecosphere,* on research conducted in Haíɫzaqv territory and citing Haíɫzaqv knowledge holders, investigates how salmon and the nutrient subsidies they bring into riparian systems impact the reproductive output of plants, focusing on salmonberries in particular. Through their work in fourteen streams, the researchers measured the impact of salmon spawning density on the reproductive output of salmonberries. Their determination is comforting in its simplicity: strong salmon runs fertilize salmon systems, the liminal spaces Haíɫzaqv think of as "salmon forests." Increased salmon density in one season leads to increased density of salmonberries per bush in the next season.

All flourishing is mutual. Thriving salmon can be read, in context, to predict thriving salmonberries, and thriving salmonberries can be read, in context, to predict thriving salmon. One key to reading the patterns lies in the kind of intimate knowledge that comes through careful observation and the tenderness of ancestral stewardship practices.

One variable missing from the paper is the role human beings have historically played in helping salmon and salmonberries to thrive. Prior to European contact, Haíɫzaqv people lived in more than fifty villages spread across more than 35,000 square kilometers of homelands on the outer central coast of British Columbia. Tribal groups and specific families were tied to salmon systems, thriving through the multilayered relationships between fish and people. Haíɫzaqv people loved systems into abundance: salmon were tended through ceremony and careful sustainable fisheries through weirs in the rivers and stone fish traps at the river mouths, while berry orchards, including salmonberry thickets, were fertilized with kelp and wood ash, crushed shells, and the blood and guts of the salmon that fed us.

As colonization decimated our populations and decades of racist laws and policies regulated us away from our homelands and ancestral practices, our ability to care for our territory was threatened. It is hardly romanticizing to say that rivers were part of our families, that those riparian systems held space for webs of kinship that were an intimate part of our existence across our territory. But the land remembers, and we are taking a deep breath in as we reassume with our full power the stewardship obligations that are written across our lands and waters.

Western science, as a practice, is not insular and unassailable. It is inherently human, a practice conducted by people who bring values and biases as a framework to everything they do. Indigenous science has its frameworks, too; though our peoples are often trivialized or romanticized as "the original environmentalists," the truth is that proper stewardship requires constant reaffirmation through the choices we make about what knowledge becomes part of our systems, how it is passed down, and how it is actualized through kincentric community-building and deep care for the lands and waters. Indigenous science is effective because our ability to mutually thrive depends on our depth of understanding of the world around us, and we choose to be guided by this science over and over each day and each season as we apply it in our stewardship work.

There is a lot that science, any iteration of science, can learn from salmon and salmonberries: as affirmed in the *Ecosphere* paper, salmon are healers, restoring the balance in the face of downstream nutrient flux from the rivers and streams as they come home to spawn and enrich the places that first nurtured them. Salmonberries patiently amplify that richness into whole thriving systems. There are patterns and stories waiting to be read and interpreted to empower wise and just standards of care for the lands and waters. And beyond the science, salmon forests and salmonberry gardens—and our plant and animal kin within them—teach us critical lessons about mutual aid and community care. We have the tools we need to flourish together.

It's March as I write this, and every day I'm out in the yard anxiously looking for new growth to give me a sense of hope. Before long, the first tight-fisted curls of vivid green leaves will appear on the skeletal salmonberry bushes outside. My children will be watching for the pink blossoms to unfurl; all through the summer and fall, they helped me bury ashes and fish guts to feed the roots of our little salmonberry orchard, and they know their demonstrations of care and reciprocity will manifest as future abundance. As the world warms and the blossoms transform into soft and brilliant orange and red fruits, my boys will take up their generational task: their ǧáǧṃ and ǧáǧṃp will sit across the kitchen table just like mine did, pour the boys' spoils into a bowl, and invite them to help intuit the salmon season ahead.

They are building patterns in their little Háíłzaqv minds, these children who were helped into the world by salmonberry leaf tonic that strengthened my womb for birth and who count salmonberries among their first foods. The stewardship pathways they are building with salmon, salmonberries, and our other nonhuman kin open them to lessons about reciprocity and interdependence that I know will inspire patience and careful observation. From that quiet place, respect and wisdom will grow. I know this because my own lessons in stewardship began with salmonberries, and because my *hbúkv* (mother) and her *q̓ísq̓* (parents), my ǧáǧṃ and ǧáǧṃp, have told me the same. And if there is one lesson I have carried forward, one lesson I hope to instill in my children, it's the importance of thriving together.

KASHMIR HILL

Your Face Is Not Your Own

FROM *The New York Times Magazine*

IN MAY 2019, an agent at the Department of Homeland Security received a trove of unsettling images. Found by Yahoo in a Syrian user's account, the photos seemed to document the sexual abuse of a young girl. One showed a man with his head reclined on a pillow, gazing directly at the camera. The man appeared to be white, with brown hair and a goatee, but it was hard to really make him out; the photo was grainy, the angle a bit oblique. The agent sent the man's face to child-crime investigators around the country in the hope that someone might recognize him.

When an investigator in New York saw the request, she ran the face through an unusual new facial-recognition app she had just started using, called Clearview AI. The team behind it had scraped the public web—social media, employment sites, YouTube, Venmo—to create a database with three billion images of people, along with links to the web pages from which the photos had come. This dwarfed the databases of other such products for law enforcement, which drew only on official photography like mug shots, driver's licenses, and passport pictures; with Clearview, it was effortless to go from a face to a Facebook account.

The app turned up an odd hit: an Instagram photo of a heavily muscled Asian man and a female fitness model, posing on a red carpet at a bodybuilding expo in Las Vegas. The suspect was neither Asian nor a woman. But upon closer inspection, you could see a white man in the background, at the edge of the photo's frame, standing behind the counter of a booth for a workout-supplements company. That was the match. On Instagram, his face would appear about half as big as your fingernail. The federal agent was astounded.

The agent contacted the supplements company and obtained the booth worker's name: Andres Rafael Viola, who turned out to be an Argentine citizen living in Las Vegas. Another investigator found Viola's Facebook account. His profile was public; browsing it, the investigator found photos of a room that matched one from the images, as well as pictures of the victim, a seven-year-old. Law-enforcement officers arrested Viola in June 2019. He later pleaded guilty to sexually assaulting a child and producing images of the abuse and was sentenced to thirty-five years in prison. (Viola's lawyer did not respond to multiple requests for comment.)

At the time, the use of Clearview in Viola's case was not made public; I learned about it recently, through court documents, interviews with law-enforcement officials and a promotional Power-Point presentation that Clearview made. The case represented the technology's first use on a child-exploitation case by Homeland Security Investigations, or HSI, which is the investigative arm of Immigrations and Customs Enforcement. (Such crimes fall under the agency because, pre-internet, so much abuse material was being sent by mail internationally.) "It was an interesting first foray into our Clearview experience," said Erin Burke, chief of HSI's Child Exploitation Investigations Unit. "There was no way we would have found that guy."

Few outside law-enforcement entities knew of Clearview's existence back then. That was by design: the government often avoids tipping off would-be criminals to cutting-edge investigative techniques, and Clearview's founders worried about the reaction to their product. Helping to catch sex abusers was clearly a worthy cause, but the company's method of doing so—hoovering up the personal photos of millions of Americans—was unprecedented and shocking. Indeed, when the public found out about Clearview last year, in a *New York Times* article I wrote, an immense backlash ensued.

Facebook, LinkedIn, Venmo, and Google sent cease-and-desist letters to the company, accusing it of violating their terms of service and demanding, to no avail, that it stop using their photos. BuzzFeed published a leaked list of Clearview users, which included not just law enforcement but major private organizations including Bank of America and the NBA. (Each says it only tested the technology and was never a client.) I discovered that the company had made the app available to investors, potential investors, and business partners, including a billionaire who used it to identify his daughter's date when the couple unexpectedly walked into a restaurant where he was dining.

Computers once performed facial recognition rather imprecisely, by identifying people's facial features and measuring the distances among them—a crude method that did not reliably result in matches. But recently, the technology has improved significantly, because of advances in artificial intelligence. AI software can analyze countless photos of people's faces and learn to make impressive predictions about which images are of the same person; the more faces it inspects, the better it gets. Clearview is deploying this approach using billions of photos from the public internet. By testing legal and ethical limits around the collection and use of those images, it has become the front-runner in the field.

After Clearview's activities came to light, Senator Ed Markey of Massachusetts wrote to the company asking that it reveal its law-enforcement customers and give Americans a way to delete themselves from Clearview's database. Officials in Canada, Britain, Australia, and the European Union investigated the company. There were bans on police use of facial recognition in parts of the United States, including Boston and Minneapolis, and state legislatures imposed restrictions on it, with Washington and Massachusetts declaring that a judge must sign off before the police run a search.

In Illinois and Texas, companies already had to obtain consent from residents to use their "faceprint," the unique pattern of their face, and after the Clearview revelations, Senators Bernie Sanders and Jeff Merkley proposed a version of Illinois's law for the whole country. California has a privacy law giving citizens control over how their data is used, and some of the state's residents invoked that provision to get Clearview to stop using their photos. (In March, California activists filed a lawsuit in state court.) Perhaps most significant, ten class-action complaints were filed against Clearview around the United States for invasion of privacy, along with lawsuits from the ACLU and Vermont's attorney general. "This is a company that got way out over its skis in an attempt to be the first with this business model," Nathan Freed Wessler, one of the ACLU lawyers who filed the organization's lawsuit, in Illinois state court, told me.

It seemed entirely possible that Clearview AI would be sued, legislated, or shamed out of existence. But that didn't happen. With no federal law prohibiting or even regulating the use of facial recognition, Clearview did not, for the most part, change its practices. Nor did it implode. While it shut down private companies' accounts, it continued to acquire government customers. Clearview's most effective sales tool, at first, was a free trial it offered to anyone with a law-enforcement-affiliated email address, along with a low, low price: You could access Clearview AI for as little as two thousand dollars per year. Most comparable vendors—whose products are not even as extensive—charged six figures. The company later hired a seasoned sales director who raised the price. "Our growth rate is crazy," Hoan Ton-That, Clearview's chief executive, said.

Clearview has now raised $17 million and, according to Pitch-Book, is valued at nearly $109 million. As of January 2020, it had been used by at least 600 law-enforcement agencies; the company says it is now up to 3,100. The army and the air force are customers. ICE signed a $224,000 deal in August; Erin Burke, of the Child Exploitation Investigations Unit, said she now supervises the deployment of Clearview AI for a variety of criminal investigations at HSI, not just child-exploitation cases. "It has revolutionized how we are able to identify and rescue children," Burke told me. "It's only going to get better, the more images that Clearview is able to scrape."

The legal threats to Clearview have begun to move through the courts, and Clearview is preparing a powerful response, invoking the First Amendment. Many civil-liberties advocates fear the company will prevail, and they are aghast at the potential consequences. One major concern is that facial-recognition technology might be too flawed for law enforcement to rely on. A federal agency called the National Institute of Standards and Technology (NIST) periodically tests the accuracy of facial-recognition algorithms voluntarily submitted by vendors; Clearview hasn't participated. In 2019, the agency found that many algorithms were less accurate in identifying people of color, meaning their use could worsen systemic bias in the criminal-justice system. In the last year, three cases have been unearthed (none involving Clearview) in which police officers arrested and briefly jailed the wrong person based on a bad facial-recognition match. All three of the wrongfully arrested were Black men.

There's also a broader reason that critics fear a court decision favoring Clearview: it could let companies track us as pervasively in the real world as they already do online.

A majority of us, members of some religious groups excepted and pandemic notwithstanding, go around showing our faces all the time. We post selfies on the internet. Walking down the street, we are unwittingly photographed by surveillance cameras and—as happened to Andres Rafael Viola—by strangers we inadvertently photo-bomb. Until recently, we've had little reason to think deeply about the fact that each of our faces is as unique as a fingerprint or a Social Security number.

Behind the scenes, though, a quiet revolution has been afoot to unlock the secrets of our faceprints. It has been powered by an enormous influx of AI expertise into Silicon Valley in recent decades, much of it drawn out of the computer-science departments of elite universities. These experts have been put to work on a number of long-term projects, including language translation and self-driving cars, and one particularly intense area of research has been facial recognition. By 2010, this effort was far enough along for Facebook to introduce a feature called "tag suggestions" that suggested the names of friends who appeared in photos uploaded to its platform. Similar features began proliferating in consumer technology: You could unlock your smartphone by looking at it and then sort all the photos on the device by face. Google's Nest camera could tell you which neighbor was at the door.

As technology advanced, policy makers didn't keep up. In the absence of robust regulations, the only thing that kept companies like Facebook and Google from going beyond those basic features we'd grown accustomed to was their own restraint. Deploying facial recognition to identify strangers had generally been seen as taboo, a dangerous technological superpower that the world wasn't ready for. It could help a creep ID you at a bar or let a stranger eavesdrop on a sensitive conversation and know the identities of those talking. It could galvanize countless name-and-shame campaigns, allow the police to identify protesters, and generally eliminate the comfort that comes from being anonymous as you move through the world.

Companies like Facebook and Google forbid "scraping," or the automated copying of data from their sites, in their terms of service. Still, by encouraging billions of people to post photos of themselves online alongside their names, tech companies provided the ingredients for such a product to succeed, were anyone audacious enough to violate the platforms' boilerplate legalese. In artificial intelligence, the more data you have, the better your product usually is. It was precisely because of Clearview's brazen collection of images from popular platforms that it was able to become its industry's leader.

The main federal law discouraging Clearview from doing that is the Computer Fraud and Abuse Act, passed by Congress in 1986, which forbids "unauthorized access" to a computer. The law was intended to prevent hacking, but some prosecutors have interpreted it as forbidding the violation of a site's terms of service, including by scraping. Clearview's executives, like many entrepreneurs who have come before them, built a company around the gamble that the rules would successfully be bent in their favor.

Their bet was partly validated in the fall of 2019, when a federal judge in the Ninth Circuit ruled in a high-profile case—which LinkedIn had filed against a start-up that was scraping its users' profiles—that automated online copying of publicly available information does not violate the anti-hacking law. The Electronic Frontier Foundation, a civil-liberties group, called the ruling "a major win for research and innovation," because it meant journalists, academics, and researchers could automatically collect information from websites without fear. But it was also an excellent precedent for Clearview and its growing database of publicly available photos. (The EFF has since called for federal protections to prevent biometric identification like what Clearview sells.)

The biggest remaining legal hurdle for the company, absent some sudden congressional action, is Illinois's Biometric Information Privacy Act (BIPA), a state law from 2008 that offers the strongest protection in the country for people's faces. The law says that private entities must receive individuals' consent to use their biometrics—a fancy word for measurements taken of the human body—or incur fines of up to $5,000 per use. In 2015, five years after introducing its facial-recognition-based photo tagging, Facebook was hit with a class-action lawsuit in Illinois for violating the law. It settled the suit last year for $650 million.

Clearview is now fighting eleven lawsuits in the state, including the one filed by the ACLU in state court. In response to the

challenges, Clearview quickly removed any photos it determined
came from Illinois, based on geographical information embedded in the files it scraped—but if that seemed on the surface like
a capitulation, it wasn't.

When I started reporting on Clearview AI in November 2019, the
company avoided me. For more than a month, its employees and
investors mostly ignored my emails and phone calls. Clearview's
then-sparse website listed a company address just a few blocks away
from the Times Building in Midtown Manhattan, so I walked over
to knock on its door—only to discover there was no building with
that address. (The company later told me it was a typo.) I had
trouble even finding out who was behind Clearview. Once the
company realized I was not going away, it hired Lisa Linden, a seasoned crisis-communications expert, to help deal with me.

In January 2020, Linden introduced me to Hoan Ton-That,
Clearview's chief executive, and we met and talked over lattes at a
WeWork in New York. Ton-That and I kept in touch. Last March,
after I told Clearview I wanted to write about how the company
was dealing with the challenges, legal and otherwise, coming its
way, he agreed to have phone calls with me every few weeks, under
the condition that I not write about them until the publication of
this article. In September, Linden invited me to observe a meeting
between Ton-That and one of the most accomplished lawyers in
the country, Floyd Abrams.

Abrams is a lion of First Amendment law, renowned for defending the *New York Times*' right to publish the Pentagon Papers fifty
years ago. Clearview had hired him, along with a national-security
lawyer, Lee Wolosky of Jenner & Block, to defend itself in the
Illinois lawsuits. Because of the pandemic, Abrams hadn't been
spending much time at the offices of Cahill Gordon & Reindel,
the corporate law firm where he is a senior counsel. So, on a sum-
mery Friday morning, Ton-That met with him instead at Abrams's
Fifth Avenue apartment in Manhattan, where visitors are greeted
by photos of Abrams shaking hands with Barack Obama and posing with Bill Clinton and George W. Bush.

In his light-filled home office, Abrams—wearing gray slacks, a
blue button-up shirt, and a black mask—sat down in a low-slung
lounge chair. Six feet away, by the window, was Linden, in a black

ensemble and floral-print mask. Ton-That walked in a minute late, dressed in a paisley jacket, a red bandanna functioning as his mask. At thirty-two, Ton-That, who has an Australian mother and claims descent from Vietnamese royalty on his father's side, is tall, slender, and elegant. With long black hair and androgynous good looks, he briefly considered a modeling career. He set his gray laptop bag on the floor and reclined in a chair that seemed too small for his lanky body. He came across as serene, without the anxiety you might expect from a person whose company was facing an existential crisis in the courts. He has a performer's ease from years of playing guitar.

Abrams immediately brought up the ACLU lawsuit in Illinois. The ACLU said Clearview had violated Illinois's prohibition on using people's faceprints without their consent. Abrams and Ton-That were working on a motion to dismiss the case, arguing that the prohibition violates the company's constitutional right to free speech.

While Floyd Abrams and the ACLU might not seem like natural enemies—the ACLU itself being known for defending the First Amendment—Abrams is embracing free speech more radically than the ACLU is comfortable with, given its concern with civil liberties other than freedom of speech, including individuals' right to privacy. In Abrams's view, Clearview is simply analyzing information in the public realm, an activity the government should not curtail. Abrams's position also reflects a career shift, from primarily defending the constitutional rights of journalists to supporting those of corporations. After the 2008 financial meltdown, he argued that AAA ratings by Standard & Poor's of debt that turned out to be junk were simply the company's opinion and therefore worthy of protection like any citizen's. He represented Mitch McConnell in the 2010 Citizens United case, in which the Supreme Court found that limiting corporations' political spending violated their free speech.

The ACLU doesn't object to Clearview's scraping of photos, but it says that creating a faceprint from them is "conduct" and not speech—and thus isn't constitutionally protected. Abrams disagrees with that and plans, he said, to argue that analyzing publicly available information (online photos, in this case) and sharing the findings (photos of one particular person) is protected by the First Amendment. Arguing that search results are speech is not without precedent: in 2003, Google won a federal case on similar grounds, after an advertising company accused Google of intentionally lowering its ranking in search results. Clearview had also gathered images from across the web and made them searchable. Google lets you search by name; Clearview lets you search by face.

Abrams saw the Google case as a useful precedent. "We're citing a case that says that a search engine's First Amendment rights would be violated if it were compelled to speak in a manner that the plaintiff wanted," he said to Ton-That. He wanted to write in the motion to dismiss that Clearview's "app makes similar judgments about what information will be most useful to its users."

Then Abrams, who is eighty-four, hesitated: "Is that the way one describes what an app does?" he asked the chief executive. "Does one say the app makes judgments?"

"I wouldn't say we make judgments but provide information," Ton-That said. Then he paused. "Well, I guess we do make judgments in what's similar, but we don't tell them it's a final judgment about who someone is."

"Provides information," Linden suggested.

"On a technical computer level, it's the computer's judgment," Ton-That added. "But we don't want that to be the final judgment when someone is arre—" He stopped himself there.

There is no documented case of Clearview's use resulting in the misidentification of a criminal suspect, but Ton-That was clearly aware that a bad match is possible. The company says that its algorithm is far superior to anything else on the market—a claim that police officers who have used it attest to—though it hasn't submitted its algorithm to NIST for accuracy testing. (Law-enforcement officers told me they would never arrest someone based on facial recognition alone and that a match is only a clue that should lead to further investigation.)

In Abrams's home office, Ton-That did a demo of Clearview. He signed into the app on Abrams's computer, then searched using a photo of Abrams. Usually results appear instantly, but there was a delay, some kind of technological hiccup. Ton-That laughed nervously. "Maybe this is less dangerous than people think," Abrams quipped. But when Ton-That searched instead for Abrams's son, Dan Abrams, a legal correspondent at ABC News, the app performed beautifully: the screen filled with a grid of photos of the younger Abrams from around the web, with the source of each identified in tiny type under the photo. Ton-That clicked on one of the photos, where he was standing with a woman, then clicked on the woman, which brought up numerous photos of her as well.

Those who support Clearview in its legal wranglings are worried that a loss would stifle innovation. "The primary goal of free speech ought to be protecting the ability to generate knowledge through mechanical means or any means," Jane Bambauer, a law professor at the University of Arizona who wrote an amicus brief in support of Clearview's position, told me.

On the other side are those who believe that a ruling in favor of Clearview's methods could usher in a future in which facial recognition is commonplace. Jameel Jaffer, a former ACLU lawyer who is now the director of the Knight First Amendment Institute at Columbia University, points out that most people who put their photos online over the last two decades very likely didn't realize their faceprint could be derived from them. He offered the example of going to a hairdresser who also collects your trimmings and sequences the DNA. "If you don't think that activity is protected by the First Amendment, you have to ask what about Clearview's activity is different," Jaffer said.

The cases against Clearview are still in early stages and will probably take years to play out. The company can continue to operate while they do. If it loses this first battle, Abrams plans to appeal, and to keep appealing as many times as needed. He predicts at least one of the cases will eventually make it to the Supreme Court, a place he has argued thirteen times in the past.

In recent cases, the Supreme Court has limited the government's use of new technologies to track people en masse, ruling that the police need a warrant, for example, to collect data about people's movement from cell phone companies. But the rights of private entities—whether individuals or companies—have been treated differently. In 2011, the Supreme Court heard a case involving a Vermont law that prohibited the sale of information about the drugs doctors were prescribing. Some companies sued, saying the law was unconstitutional because they had a free-speech right to buy and sell that information. The Supreme Court ruled in favor of the companies.

Clearview and the ACLU will appear before a judge in April to discuss the motion to dismiss. The fact that Clearview's database is made up of public photos is the core of Abrams's defense.

"We're saying that where information is already out, already public," Abrams said, "that the First Amendment provides enormous protection."

During the year I've been reporting on Clearview, one mysterious subject has been the exact details of the company's origins. According to Ton-That's version of events, he and a man named Richard J. Schwartz founded Clearview AI together. But the pair always struck me as an odd match. Ton-That moved to San Francisco from Canberra in 2007 at age nineteen to chase the tech gold rush, spinning up moderately successful Facebook games and iPhone apps and attending Burning Man, but then eventually decamped for New York in 2016. Schwartz is a grizzled New York politico who worked for Mayor Rudolph W. Giuliani in the nineties, edited the *New York Daily News* editorial page, and did communications consulting. He is thirty years older than Ton-That and seems to come from an entirely different world. So, last year, I asked Ton-That how they met and came to found the company together.

Ton-That said he encountered Schwartz in 2016 at the Manhattan Institute, a conservative think tank, during a book event. He said they talked for an hour and decided to meet again for coffee the following week. That time, they chatted for three hours, including about technology and public policy. "And it went from there," he said. Schwartz later told me he was intrigued by the idea of joining Ton-That's "brilliant mind and exceptional technical skills with my experience, relationships and know-how." When the company was first registered in New York in February 2017, using Schwartz's apartment on the Upper West Side as its business address, it was called Smartcheckr LLC. The name changed to Clearview AI the following year. Ton-That was vague about what happened in those early years, declining to name others involved beyond Schwartz. In Ton-That's telling, the company just kind of stumbled into facial recognition.

That story never satisfied me. Clearview is a radical new entrant to the technological scene. It dared to contravene a taboo that Google and Facebook—not generally known for their privacy-respecting ways—saw as exceedingly unwise to cross. For the last year, I have tried to figure out the exact genesis of that iconoclastic development and learned that the company's origin story is more complex than Ton-That made it out to be.

After I broke the news about Clearview AI, BuzzFeed and the *Huffington Post* reported that Ton-That and his company had ties to the far right and to a notorious conservative provocateur named Charles Johnson. I heard the same about Johnson from multiple sources. So, I emailed him. At first, he was hesitant to talk to me, insisting he would do so only off the record, because he was still frustrated about the last time he talked to a *New York Times* journalist, when the media columnist David Carr profiled him in 2014.

Back then, Johnson was a twenty-six-year-old blogger who would try to poke holes in big stories that were popular with progressives. When a police officer killed eighteen-year-old Michael Brown in Ferguson, Missouri, Johnson sued unsuccessfully to obtain Brown's juvenile records and published photos from Brown's Instagram account that he claimed showed a violent streak. Later, *Rolling Stone* wrote about a University of Virginia student named Jackie who claimed that she was gang-raped at a fraternity, and Johnson called the story a hoax; after the magazine acknowledged discrepancies in Jackie's story, Johnson posted what he said was her last name, along with photos of her. *Rolling Stone* later retracted the story altogether. Carr criticized Johnson's attack-dog tactics and noted factual errors, calling Johnson a "troll on steroids," but pointed out that he had gotten some notable scoops and was "not without some talent."

Johnson found his tactics and political leanings suddenly becoming more mainstream during the Trump administration, and he began to accumulate real influence. *Forbes* reported that he helped the White House vet political appointees.

Johnson says he eventually decided to talk to me on the record because he regrets some of his decisions and the notoriety that has haunted him since. He wanted to correct what he feels are mistaken impressions of him by revealing that he helped start a company whose product is now being used to save children from sexual abuse.

Johnson claims that he met Ton-That in 2016, introduced him to Schwartz, and considers himself a third cofounder of Clearview. I was skeptical at first, given Johnson's reputation as a peddler of disinformation. In a statement, Ton-That acknowledged that he met Johnson in 2016 and that Johnson had "introduced people to the company." But he said Johnson was not a founder and never had an operational role. Johnson, however, provided email and legal documents that, along with other sources, strongly support his claims; indeed, the company might not exist without his contributions.

According to Johnson's version of events, which Clearview disputes, it all began in May 2016, when Ton-That emailed Johnson, saying he was an admirer of Johnson's work and asking to join a Slack group that he ran for fans of his right-wing takes. The next month, Johnson visited New York, and Ton-That met him for the first time in person. They hung out for at least ten hours straight and became fast friends, according to Johnson and associates of Ton-That at the time. The people who knew Ton-That said he had always been contrarian, but it surprised them when he came out as a Trump supporter in early 2016. They worried about his new relationship with Johnson, given his extreme views and associations. Ton-That recently described himself as "confused" at that time in his life. He went on: "People get radicalized into things. It's crazy to see it. I got sucked in for a while."

That summer, the new friends attended the Republican National Convention in Cleveland, where Donald Trump was being crowned the party's presidential nominee. Johnson had rented a big group house on Airbnb. "Am I still allowed to crash?" Ton-That wrote in an email to Johnson, which Johnson provided to me. "I'll bring my guitar, can chip in for accommodations."

"Yes, of course," Johnson replied. "Want to meet Thiel?"

"Of course!" Ton-That wrote back.

"Thiel," of course, was Peter Thiel, one of the most powerful men in Silicon Valley—though he no longer lives there, having moved to Los Angeles. (A spokesman for Thiel did not respond to requests for comment.) He famously turned an early $500,000 investment in Facebook into a billion dollars and became a founder of Palantir, a data-gathering juggernaut.

Thiel was in Cleveland because he had come out in support of Trump and was giving a prime-time speech at the convention. Johnson sent me a photo taken of him and Ton-That on the floor of the arena: both men are smiling, with Thiel visible on a screen behind them.

While Johnson and Ton-That hung out at the rental house, they mused about discredited sciences that could be explored in the modern age with new technologies. At one point, the conversation turned to physiognomy, the pseudoscientific judgment of a person's character based on their facial features. "Hoan played music," Johnson said. "We all drank a lot." He added, "That was where a lot of ideas that became Smartcheckr, and then Clearview, began." Johnson told me he also arranged a meeting between Thiel and Ton-That at a home in Shaker Heights that week.

Johnson says he was the one who brought in Schwartz, because of Schwartz's deep political connections in New York—including at the NYPD—and because he offered an inroad to Trump as a former Giuliani lieutenant. Two days after the convention ended, Johnson emailed Ton-That and Schwartz, introducing them. Within a week, they made plans to meet, according to an email thread that Johnson forwarded to me.

Seven months later, in February 2017, Schwartz emailed draft formation documents for a company called Smartcheckr LLC to Johnson, which granted equal ownership to Schwartz, Ton-That, and Johnson. It was a name that would seem to have Johnson's fingerprints all over it—he previously founded start-ups called WeSearchr and FreeStartr—though the company claims the name was Schwartz's idea. "I am very excited about our new company and look forward to the great work you, Hoan and I will be doing together!" Schwartz wrote.

Ton-That says the LLC "was not intended for the purpose of developing facial-recognition technology, and it conducted no business." Johnson claims the plan from the beginning was to make an app to identify faces. In June 2017, Ton-That emailed Schwartz, Johnson, and another person a link to a *Scientific American* article about Caltech researchers who had shed new light on how the brain identifies faces. Schwartz responded, "Sounds like Caltech is a year behind you."

In July 2017, a director at Thiel Capital, an investment firm founded by Thiel, emailed Ton-That to say that Thiel was interested in investing $200,000. Ton-That forwarded the email to Johnson. Thiel soon did invest.

Johnson was living on the West Coast, dealing with a new child and a disintegrating marriage, and while he was introducing the company to potential funders and clients, he was not involved in day-to-day operations. In August 2017, Smartcheckr registered as a corporation in Delaware. This time, Schwartz and Ton-That were listed as the only directors.

That fall, perhaps trying to keep some money coming in while improving its facial-recognition technology, Smartcheckr pitched itself to political candidates as a consulting firm. A person close to the company in its early days said the founders wanted to dig up dirt on liberals, which the company and Johnson deny. Paul Nehlen, a far-right Republican running for Congress in Wisconsin, publicly claimed the company had sent him a brochure about "enriched" voter profiles, "microtargeting" of voters, and "extreme opposition research." (Nehlen didn't respond to requests for comment.) When I asked the company about his claims last year, it told me it never actually offered such services and that the email came from a rogue contractor. But I found out that it was not a one-off—nor was the outreach limited to Republicans.

Schwartz offered the same Smartcheckr services, in October 2017, to a Democratic newcomer to politics named Holly Lynch, a communications consultant who was running for a congressional seat in New York. According to Lynch, he told her he had a great guy who could be very helpful with voter data, called the Prince—a reference to Ton-That's royal ancestry. Lynch said Schwartz didn't mention facial recognition, only "unconventional databases." Lynch ultimately chose not to work with Smartcheckr and soon ended her campaign.

It appears Smartcheckr decided against pursuing political consulting. The facial recognition it had been working on had improved. "It wasn't clear it would work until April 2018, when the accuracy part really kicked in," Ton-That said.

Two months later, the company changed its name to Clearview AI. That summer, it pitched itself as a security start-up and conducted pilot facial-recognition projects with branches of TD Bank and Gristedes Supermarket in Manhattan, according to a document provided to a potential investor. (Gristedes's owner, John Catsimatidis, confirmed its project; TD Bank said it "does not have a business relationship with Clearview AI and does not use any of Clearview AI's products.") Another investor who was approached by the company said that the product was impressive but that the ties to Charles Johnson scared him off. (He did not want to be named, fearing retribution from Johnson.)

During the course of 2018, Clearview's database grew to a billion faces from twenty million. At the end of the year, the founders dissolved the LLC they formed in New York and asked Johnson to sign a "wind-down and transfer agreement," which converted his one-third ownership in Smartcheckr LLC into a 10 percent stake in Clearview AI. The contract also entitled him to a 10 percent sales commission on any customers he introduced to the company, though Johnson hasn't been paid a commission.

The wind-down agreement, which Johnson provided to me, requires him not to "publicly disclose the existence of this agreement, his indirect ownership of the shares or his prior provision of services to the company." It is signed by Johnson, Ton-That, and Schwartz. (In early March, Clearview amended its incorporation documents such that any shareholder who "breaches any confidentiality obligations" can have his or her shares bought back at 20 percent of market value. When I told Johnson about this, he responded, "That's probably not good for me.")

Johnson said in February that he was willing to break the agreement, both because he's upset about having been erased from Clearview's past and because he thinks the company should have gone further than it has in making the technology available. Johnson believes that giving this superpower only to the police is frightening—that it should be offered to anyone who would use it for good. In his mind, a world without strangers would be a friendlier, nicer world, because all people would be accountable for their actions.

"I think Clearview should be in the hands of the moms of America," he said.

No matter its parentage, Clearview was inevitable. All the building blocks were there; it was just a matter of picking them up and putting them together. But it makes sense that Thiel, who seems to see personal data as a resource to be mined for riches, and Johnson, who made a career of digging up dirt on people, were part of the company's origins. Our faces are crucial to linking the digital data that's been accumulated about us with our identities in the real world. That is valuable not just to law enforcement but also to companies, advertisers, journalists, and, yes, the moms of America.

The fact that this superpower is not yet available to us all may just be a fluke of history. Suppose it had been Charles Johnson, not Hoan Ton-That, who ended up at the company's helm. Or

suppose—even before Clearview began—that an influential executive at Google or Facebook had successfully pushed for using the
photos and algorithms they already had to let people search for
faces as easily as we now search for text.

In some countries, facial recognition is already becoming as
mainstream as other once-unimaginable technologies now taken
for granted. In 2016, a Russian company called NTechLab developed a facial-recognition algorithm used in an app called FindFace,
which matched photos of strangers to profiles on VK—essentially
Russia's Facebook. Within months of its release, it was reported
that people were using the app to identify sex workers, porn stars,
and protesters. NTechLab shut down the public FindFace app but
still provides its algorithm to governments and corporations. In
2019, the technology was placed in Moscow surveillance cameras,
providing a live log of who passed the cameras and when. Meant
to be used to find criminal suspects, it was repurposed to enforce
lockdown during the COVID-19 pandemic. In March, a man who
was supposed to be quarantining went outside his apartment to
take out the trash; thirty minutes later, the police were at his door.

In China, facial recognition aids in surveilling the population
and in enforcing both the law and social norms. In Suzhou, local
authorities have deployed it to name and shame people wearing
their pajamas in public. Other uses are quite a bit more sinister,
including automatically flagging the faces of Uighurs and other
ethnic minorities and tracking their comings and goings. In 2018,
Chinese police officers began testing out facial-recognition glasses
that would let them more easily ID the people they interact with.
When the *New York Times* analyzed a copy of the computer code
underlying the Clearview AI app, a data journalist at the paper
found that it, too, was designed to be able to run on augmented-
reality glasses. (The company says it has experimented with this
function only in its lab.)

Facial recognition would of course look different in the
American context, where the state's reach is significantly more
curtailed—by both laws and norms—than it is in China or Russia. The more society-changing aspect of facial recognition in the
United States may be how private companies deploy it: Americans'
right to privacy is relatively strong when it comes to the federal
government but very weak when it comes to what corporations
can do. While Clearview has said it doesn't want to make its app

available to the public, a copycat company could. Facebook has already discussed putting facial recognition into augmented-reality glasses. Within the last year, a mysterious new site called PimEyes has popped up with a face search that works surprisingly well.

Retail chains that get their hands on technology like this could try to use it to more effectively blacklist shoplifters, a use Rite Aid has already piloted (but abandoned). In recent years, surveillance companies casually rolled out automated license-plate readers that track cars' locations, which are frequently used to solve crimes; such companies could easily add face reading as a feature. The advertising industry that tracks your every movement online would be able to do so in the real world: that scene from *Minority Report* in which Tom Cruise's character flees through a shopping mall of targeted pop-up ads—"John Anderton, you could use a Guinness right about now!"—could be our future.

And imagine what you would do with a face-identifying app on your phone: a Shazam for people. You would never forget someone's name at a party again. If that pseudonymous troll on Twitter who said something nasty to you had ever tweeted a selfie, you could find out who he or she was. You could take a photo of the strangers at your poker table and know if they're pros or not. It might just be your new favorite app.

Alvaro Bedoya, a former congressional staff member who started a privacy center at Georgetown Law School, told me widespread facial recognition could both empower the government and transform civilian life—outcomes that he called "equally pernicious." He thinks, for example, that ICE could start searching out visa overstayers for deportation by using the photos taken when they entered the country and scanning surveillance-camera feeds for them once their documentation expires. And anonymity could be eradicated in day-to-day life.

"When we interact with people on the street, there's a certain level of respect accorded to strangers," Bedoya told me. "That's partly because we don't know if people are powerful or influential or we could get in trouble for treating them poorly. I don't know what happens in a world where you see someone in the street and immediately know where they work, where they went to school, if they have a criminal record, what their credit score is. I don't know how society changes, but I don't think it changes for the better."

*

It's impossible, of course, to perfectly predict how novel technologies will ultimately be used and how they will reshape our world. On the day the Capitol was stormed by pro-Trump rioters in January, Ton-That was at work in his Chelsea apartment. Then his phone began to buzz with text messages and phone calls from friends and colleagues, predicting that Clearview AI would be critical for identifying participants; despite the pandemic and the seemingly obvious incentives to conceal their identities, most of the rioters' faces were exposed. One of Ton-That's salespeople called because a police officer wanted free access. "I said we could because it was an emergency situation," Ton-That said.

And in fact, the next day, the company saw a surge in searches from law enforcement. The FBI wouldn't discuss whether Clearview AI was being used for its investigation of the riot, but detectives in Alabama and Florida who collaborate with the bureau at real-time crime centers said they had identified possible rioters using Clearview and sent them to the FBI. "We are up to six potential matches," an assistant Miami Police Department chief, Armando R. Aguilar, told me a week after the riot. The following week, the number was thirteen.

It was a remarkable turn of events. The relationships behind Clearview had germinated at an event celebrating Trump, at least according to Johnson; now, four years later, the app was being deployed in a domestic crackdown on lawbreaking Trump supporters. There had been a time when public opinion seemed set firmly against facial recognition. But suddenly—with people showing their faces while rampaging through the Capitol—Clearview and similar products seemed quite appealing.

Ton-That and I talked on the phone just a couple of days after the riot. He sounded tired and spoke hurriedly—he was pressed for time, he said, because of the incoming demand from law enforcement. He didn't seem to harbor any remaining allegiance to Trump, calling the attack "tragic and appalling" and declaring that the transition of power should be peaceful. While he was clearly taken aback by the events unfolding in his adopted country, he also seemed keenly aware it could demonstrate the utility of his company's product, and perhaps sway those on the fence if it played a role in finding and punishing the people involved.

"You see a lot of detractors change their mind for a somewhat different use case," he said. "We're slowly winning people over."

RUTH ROBERTSON

Quantum Enlightenment

FROM *Atmos*

FOR MANY, Western science has become rather fixed and station-ary. It's based on hard data and described in absolutes. Scientific laws and theories, especially in academia, are upheld as established categorical certainties and alternative notions are often scoffed at and viewed with derision.

What you will find, however, is that the most complex spheres of science are not concrete at all.

Quantum mechanics is a body of scientific laws that mathemat-ically explain what classical physics cannot—namely, the proper-ties and behavior of subatomic particles. What's fascinating about quantum mechanics is that it's a relatively young branch of science, and so much of what's proven thus far flies in the face of what the classical sciences have accepted as gospel truths. In fact, quantum mechanics often runs contrary to logic.

This new area of science is not about absolutes. It's about prob-abilities and possibilities. And it's forcing experts to expand their definition of reality and embrace the unknown.

For example, though confounding, experiments have verified that the very nature of the subatomic world seems to be affected by whether or not we are observing it. Without us, these miniscule twinkles of existence operate beyond our current understanding. What makes them recognizable, measurable, and indeed, "real," is our watchful eye.

Attempting to explain nature on the atomic level is leading hu-man intellect into uncharted territory. String theory has become a prominent means of unifying quantum mechanics with classical physics like the law of gravity. The thing about string theory, though, is that it requires at least ten dimensions for the math to work. Humanity is presently operating within four dimensions. These alternate dimensions ultimately translate to the existence of an infinite number of parallel universes, which may have different physical laws from our own.

Quantum physicists call it parallel dimensions and the multiverse. My Native ancestors call it the spirit world.

Overall, scientists have been hesitant to investigate the spiritual since the discipline's inception, but quantum mechanics made great strides thanks to those who were willing to buck the secular constraints of university circles. Fundamental Fysiks Group was composed of underfunded physicists at Berkeley in the 1970s who used Eastern mysticism to delve deeper into quantum theory. Their work is now considered foundational to quantum mechanics.

David Bohm was a renowned quantum physicist who figured out wave-particle duality. Subatomic particles actually don't behave as expected at all. They defy reason by having the characteristics of both waves and particles. While other physicists used uncertainty to describe this behavior, he defined it as inherent ambiguity. Bohm coined the idea of a pilot wave, clarifying that it was our efforts to observe particles that changes their behavior by disturbing a pilot wave composed of subatomic particles.

What Bohm failed to dispose of was the problem of a subatomic phenomenon called nonlocality. Simply stated, it's the capacity of a particle to influence another instantaneously across great distances seemingly without explanation. Einstein referred to nonlocality as "spooky action at a distance."

Bohm once again defied convention by proposing an unusual answer to explain nonlocality: that science will never fully explain the world. In 1980, he wrote a book on it called *Wholeness and the Implicate Order* where he broached the subject of spirituality and deduced that complete clarity is not within humanity's grasp.

In his book, Bohm said that under physical appearances, or the explicate order, there is a deeper, hidden implicate order. He first visualized this arrangement by studying subatomic particles and developing the supposition that they abide in a field consisting of an infinite number of fluctuating pilot waves. From that assumption, he discussed the possibility that space and time itself might be manifestations of a deeper, implicate order.

But he didn't stop there. Bohm was primarily concerned with enlightenment. He thought that today's physicists should consider wiping the slate clean just as Newton and Descartes had done with the ancients.

It's not that we should dispose of hard math and mechanics altogether. It's just that in order for science to jump forward, it must merge with art and spirituality. He understood the importance of thinking and perceiving differently.

A key underlying concept that we shouldn't miss here is that in order for humanity to progress intellectually, or otherwise, we must become comfortable with not knowing. This does not mean that we should not strive to learn all that we can about our Universe. Instead, it says that there is power in the acceptance that we may never fully know everything because, like the many dimensions and universes that surround and overlap our own, knowledge is infinite, and in our corporeal forms, may be beyond our understanding.

My People, the Oceti Sakowin (Great Sioux Nation), acknowledge the authority of Takuskanskan, a deity that we can only describe in English as being "that which moves all things." It is the Great Mystery, but also the Universe, and everything within it. It is inside you and me. It is the Source of all life, but also motion absent any living thing. It is both the pilot wave and as Einstein said, the "spooky action at a distance." Takuskanskan is the knowledge of an observable phenomena, without being able to fully explain what it is. Takuskanskan is the humility to accept the impossible as real, and also the understanding that our reality is determined by our interaction with our surroundings.

Quantum theory and the study of the subatomic realm has also led to the suggestion that the destruction of Earth could spell the annihilation of our universe.

It is these possibilities that will determine how far humanity will evolve. In order for science and the human race in general to advance, we must dispense with the limits we have imposed on ourselves. These limitations are often related to faulty prejudices created by a system built to uphold the bogus social hierarchy of colonialism, capitalism, heteropatriarchy, and white supremacy. It is crucial that Indigenous ways of knowing are included in the scientific process because they are the antithesis of these limitations. Incorporating them will foster enlightenment and open the door to the merger necessary for greater understanding.

Futures We Could Have

CHRIS MALLOY

Why Combining Farms and Solar Panels Could Transform How We Produce Both Food and Energy

FROM The Counter

ON THE OUTSKIRTS of the famed research facility Biosphere 2, away from the futuristic glass pyramid and tiered trapezoids alive with self-contained habitats, University of Arizona biogeographer Greg Barron-Gafford oversees a small outdoor garden. It's a darker, cooler plot than most, thanks to its distinctive feature: solar panels. Mounted on nine-foot-tall beams, so lofty a tractor could pass below them, the solar panels slant against the blue sky of the high desert, throwing shade on the short rows of basil and onions beneath.

Barron-Gafford has been testing agrivoltaics—a term for land that combines agriculture and solar farming—for eight years. He started with a single solar panel at Biosphere 2, in Oracle, Arizona, a site the University of Arizona has owned since 2011. More recently, his project has expanded to sites in nearby Tucson and even a large plot overseen jointly by the National Renewable Energy Laboratory (NREL) in Longmont, Colorado.

"Making renewable energy from solar isn't that complicated anymore," said Barron-Gafford, crouching by a shadow-thatched row of carrots to check soil moisture, to be compared to a nearby control garden without an overhead solar system. "The price has really come down. The question is, where do you put it?"

It's a good question. In order to address climate change, President Joe Biden campaigned on a plan for the United States to move to clean electricity by 2035. In 2019, 11 percent of U.S. energy came from renewable sources. Just some 2 percent was solar. To reach the 2035 goal, solar will very likely need a much bigger slice of our energy portfolio.

That, though, will require expansion over land. Like windmills, solar panels need space, creating what has been dubbed "energy sprawl." Even as a fraction of the U.S. energy portfolio, solar power has already led to land-use conflicts, with proponents of solar starting to spar with farmers over land.

Agrivoltaics help to solve that spatial dilemma. They allow a given area to harvest the sun not only once, but twice—as fuel for crops and as a source of renewable energy. But space-saving isn't all that has sparked the interest of researchers and advocates across the world, from the United States to Western Europe and Japan. Another kind of symbiosis can occur, with surprising benefits we are only just beginning to understand—an untapped synergy that may have the potential to transform the way we produce both food and energy.

For Barron-Gafford, the idea emerged organically from a challenge in his work. In 2012, he was looking for ways to quantify the temperature of large solar arrays, trying to determine how they might contribute to an urban heat island effect, or the phenomenon of cities being warmer than the surrounding countryside. He noticed that heat collected beneath the solar array, making the panels less efficient. Maybe, he thought, growing plants under the panels would cool them. Plants, after all, "exhale" water through a process called transpiration, losing water through their pores as they intake carbon dioxide for photosynthesis.

Indeed, Barron-Gafford's agrivoltaic garden proved to produce a cooler microclimate under panels than did conventional solar farms, something other studies have echoed with findings from their own plots. Intrigued, he went on to research the phenomenon in more detail. In his 2019 study of jalapeños, cherry tomatoes, and chiltepines (a desert chile native to the United States), Barron-Gafford found that agrivoltaic panels produced 3 percent more energy during the May-through-June growing season and 1 percent more overall.

But the benefits weren't only on the energy-efficiency side. His simple system—just solar panels lofted above plants—also helped the crops, which needed far less water to grow. This was due to cool shade under the solar panels. "If you spilled your water bottle in the shade versus out in the sun, where's it going to stay wet longer? In the shade," he said. "So, we're just using these *oh duh* principles to try to make a more sustainable food system."

But the shade also conferred another significant benefit, one that may prove tantalizing for farmers focused on yield: it caused Barron-Gafford's plants to grow differently.

Some plants thrive in partial sun, especially in semiarid regions where farms may be growing non-native crops unadapted to water scarcity, like the Southwest. That makes solar panels an ideal growing partner. The shade they provide keeps moisture in the ground longer, which keeps plants from stressing out, Barron-Gafford explained; instead of burning through their energy reserves early, crops can keep photosynthesizing at a more even, measured rate throughout the day.

In some species, that leads to markedly increased production, Barron-Gafford found: The presence of plants ups solar energy production, and solar panels returned the favor for two of the three plants he studied. Tomato production doubled, with 65 percent more water efficiency. Chiltepin production tripled. Jalapeño production was static, yet with water efficiency 157 percent greater— a win in a region embroiled in historic drought.

While those numbers represent astonishing productivity gains, similar results may not be possible in every region or with every crop, said Jordan Macknick, lead energy-water-land analyst for NREL. Macknick has studied agrivoltaics with NREL since 2010, working to understand "what grows under solar panels where and why." Today, he analyzes data from twenty-five sites, including the plot at Biosphere 2, together forming "a network of projects that are designed to explore the tradeoff between solar energy development and agriculture."

It's clear that what works in Arizona might not work in the Pacific Northwest, and what works in New England might not work in the South. "One of the challenges is many things in agrivoltaics are very site specific," Macknick said. "You have to try very hard to be able to extrapolate results from one region to another region." His research partners take a consistent approach to enable comparison.

Agrivoltaics projects across the world already demonstrate some of this inherent variability. In Minnesota, solar panels shading the cattle of a rotational grazing pasture have successfully lowered animal heat stress while powering a milking parlor. In Belgium, an agrivoltaic potato field showed cooler microclimates below panels and less evaporation, as well as larger leaves with "adapted light harvesting capability." But in other locations, the results are more mixed. In Massachusetts, numerous cranberry bog farmers have adopted agrivoltaics, hoping that solar energy will provide an additional source of revenue and offset plummeting prices—though some worry that the crop won't respond as well to shade, the Associated Press reports. In Italy, spinach yields decreased in one study. Still, its researchers observed more efficient photosynthesis and greater protein production. They calculated that the financial value of the agrivoltaic spinach remained greater, given the solar energy production.

Generally, research suggests that agrivoltaics might benefit, at the very least, a small fraction of global farmland. But agrivoltaics won't need to work everywhere to become a transformative idea. Writing in the journal *Scientific Reports*, scientists from Oregon State University found that if a mere 1 percent of global cropland gained solar panels, "global energy demand would be offset by solar production."

Think about that for a second: turning just 1 percent of cropland agrivoltaic—a method that can also increase harvests in many environments—could more than satisfy the world's energy demand, generating BTUs by the hundred quadrillion.

Though sensible in theory and simple in imagination, turning visions of solar-paneled farms into reality comes with its own challenges. Agrivoltaic systems are no panacea: they can introduce new uncertainties, exacerbate old complexities, and many questions remain.

First, there is the question of farming logistics. Crops that require aerial spraying, like cotton, likely won't thrive in agrivoltaic systems. There have been concerns about how rain might run from panels and cause erosion, though these concerns were unfounded in both a German study and in Barron-Gafford's work. Also, low panels could obstruct farm vehicles. Barron-Gafford and other experimenters tend to keep panels six to nine feet off the ground—too low for some large combines, but high enough for many tractors. These vehicles can raise dust onto panels, cutting efficiency. A study notes that self-cleaning panels would help, and so would tweaking irrigation to clear panels before dripping to crops, "thus facilitating effective water usage."

Second, crop constraints loom. Some plants dislike partial sun. Corn, for instance, has shown weaker stalk growth and lower photosynthesis rates in shade. Scientists like Barron-Gafford are still determining which crops do best under panels.

More broadly, agrivoltaics introduce new variables to the business of farming, which is already chock-full of them. One undesirable factor: lightning can strike solar panels. A study anticipates these strikes will happen about once every thirty years and can be averted via lightning rods.

Other challenges are more vexing. In a survey conducted by Michigan Technological University, farmers expressed concerns over not knowing how crop or livestock markets might look in the years ahead. A farmer who makes an investment in agrivoltaics must look ten, twenty, thirty years into the future to project revenue, and calculating profit margins is still a speculative exercise.

Additionally, some farmers surveyed wondered how the concrete and beams might change the land over time, with an influx of large, permanent structures potentially causing issues for crops or for livestock.

Perhaps most significant, the up-front costs of agrivoltaics are giant. They will have to be defrayed through some combination of policy, tax incentives, favorable loans, and/or integration into electrical grids, where farmers can profit from selling the sunlight they farm. In the end, the balance of up-front capital costs for installation, solar energy sales, and agriculture sales must exceed the sales of a traditional farm.

These novel challenges have led to some unusual partnerships and approaches. Solar grazing, for instance, typically means bringing in livestock from outside farms and ranches to help maintain a solar site.

"With the solar . . . the animals have shade in abundance," said Lexie Hain, executive director of the American Solar Grazing Association, who practices solar grazing with sheep in upstate New

York. "If I visit my sheep at lunchtime, under the panels they're asleep, they're chewing their cud, they are not stressed. And they drink less water than they would if I had them home here on the farm."

Like many solar grazers, Hahn transports her sheep to solar farms to graze. The owners need vegetation trimmed so it doesn't block or damage panels—so, for a price and feed for their animals, solar grazers step in. "I can do it for about half of what the landscapers do," said Richard Cocke, a shepherd who brings his sheep to two solar sites in southern Arizona, having stopped visiting a third due to coyote troubles. "I don't cause breakage to the panels. And it's green."

Pollinator-friendly sites are another subcategory. According to Yale research, this approach—which entails growing beneficial nonfood plants like perennial wildflowers and native grasses under solar cells—can increase groundwater recharge, curb soil erosion, and up solar efficiency by creating a cooler microclimate for panels. It can also bolster crop yields on nearby farms of crops requiring pollination. But large-scale adoption, the study authors write, will "warrant policy intervention." With no crop to sell, pollinator-focused solar farmers would likely need lawmakers and business leaders to recognize the ecosystem services they provide, ultimately developing systems of financial incentives to make the extra trouble worth it.

The need for supportive policy is echoed down the chain of people involved in agrivoltaics, including by Byron Kominek, the owner of Jack's Solar Garden, a site with which Macknick and Barron-Gafford work. This Longmont, Colorado, agrivoltaic plot is the largest in America.

Completed in late 2020, the 3,200-panel solar system at Jack's Solar Garden spans five acres and can power more than three hundred homes. Previously, the farm produced hay. Now, Kominek will grow vegetables and whatever researchers want to test, largely for data collection. "Maybe after some years, after our researchers are done studying here . . . we'll have a sense of what we can do with the land underneath the solar panels and [can] start growing," Kominek said.

To start, Jack's makes money by selling energy. It has subscribers that, through a local energy company, buy power from its solar grid almost like you would produce from a CSA. The biggest subscriber is a local cannabis company. These cash infusions have helped with the "big expense" of installation. "We were able to cover it because we basically asked our subscribers to pay us upfront for X number of years," Kominek said. "This helps us have the up-front capital that we can pay off our engineers who helped build the system."

Kominek hopes that legislators will look at Barron-Gafford's projects and other developments, like the budding scene in Massachusetts, the only state with a financial incentive for agrivoltaics (six cents per kilowatt-hour). He hopes the practice can be fostered via rulemaking.

"In the end, these things only pan out if the finance is there," he said. "They don't work on hopes and dreams."

At sites like the agrivoltaic garden on Biosphere 2, a complex built to test how people might live on other planets, we are still learning clues for living on this one.

Barron-Gafford's young winter crop includes six plants, like chickpeas and fava beans, which will also appear in a sister garden run by The Arava Institute for Environmental Studies in the Arava Desert of Jordan and Israel, home to a climate similar to the desert of southern Arizona. Increasingly, agrivoltaics researchers are collaborating to suss out the nuances of the technology, which—though expensive, not for every climate or crop, and arguably an aesthetic blight on the land—packs layers of potential.

Standing with his boots in the dirt of the American county with the second-most climate risk, Barron-Gafford ponders the future ahead. "What happens in another fifty years when we're under increased pressure because of dropping reservoirs?" he asks. "What about fifty years when our temperatures are beyond what plants can handle?"

For the United States to meet its electricity goals and to avert greater climate chaos, renewable energy will likely have to grow. So, the fact that the world could meet energy demand by adding solar panels to 1 percent of cropland suggests agrivoltaics are a conundrum worth puzzling over. With the right policies, incentives, and farmer support, the technology might be a rare win-win. It might even be low-hanging fruit.

SARAH KAPLAN

A Recipe for Fighting Climate Change and Feeding the World

FROM *The Washington Post*

"IT'S SO DIFFERENT from anything I've baked with," says my baking partner, Jenny Starrs.

We're standing in the tiny kitchen of my D.C. apartment, examining palmfuls of a dark, coarse, rich-scented flour. It's unfamiliar because it was milled from Kernza, a grain that is fundamentally unlike all other wheat humans grow.

Most commercial crops are annual. They provide only one harvest and must be replanted every year. Growing these foods on an industrial scale usually takes huge amounts of water, fertilizer, and energy, making agriculture a major source of carbon and other pollutants. Scientists say this style of farming has imperiled Earth's soils, destroyed vital habitats, and contributed to the dangerous warming of our world.

But Kernza—a domesticated form of wheatgrass developed by scientists at the nonprofit Land Institute—is perennial. A single seed will grow into a plant that provides grain year after year after year. It forms deep roots that store carbon in the soil and prevent erosion. It can be planted alongside other crops to reduce the need for fertilizer and provide habitat for wildlife.

In short, proponents say, it can mimic the way a natural ecosystem works—potentially transforming farming from a cause of environmental degradation into a solution to the planet's biggest crises.

This summer I traveled to Kansas, where I met the scientists who are trying to make Kernza as hardy and fertile as traditional wheat. I visited the farmers who must figure out how to grow it effectively. And I invited my friend Jenny, the founder of artisan baking company Starrs Sourdough, to help me make a loaf of Kernza bread.

Kernza has a long road from the laboratory to the kitchen table. It will be even harder to transform the farming practices that humans have relied on for most of history. But if the scientists, farmers, and processors are successful, perennial foods might one day be available on grocery store shelves—and the bread that Jenny and I are baking could offer a taste of what's to come.

The Soil

The first step in Jenny's bread recipe is making the "levain"—a mix of flour, water, and yeast that ferments for a long time, producing lots of air bubbles and tasty lactic acid.

While the microbes chow down, Jenny and I compare the whole Kernza to some wheat kernels she has on hand. The Kernza grains are smaller, and they contain less of the gluten protein that makes traditional wheat good for baking bread.

"Obviously, bread flour is awesome," Jenny says—after all, humans have been perfecting it for nearly ten thousand years.

At the end of the last ice age, in the fertile river valleys of the Middle East, China, and Mexico, people found they could sustain themselves more easily by cultivating crops. Three annual grasses—wheat, rice, and corn—became the foundation of human diets and human civilization.

Freed from the need to rove the landscape in search of food, people settled down and constructed cities. Religions and school calendars were structured around the rhythms of farming: planting seeds, helping them grow, harvesting grains, and then tilling the soil to prepare it for the next round of planting. Generations of careful breeding improved crops' taste and yield, and ever-stronger fertilizers have made farms still more productive. The population boomed.

But the planet has paid the price. The practice of tillage—churning the ground to destroy weeds and facilitate the planting of next year's crop—has depleted the very earth from which our food is grown. It breaks up clumps of organic matter and exposes them to the sunlight, releasing carbon into the atmosphere. Tilled soil is less able to hold water, causing nutrients and other particles to run off into rivers, lakes, and the sea.

Research suggests that the world's soils are now eroding a hundred times faster than new soil can form, and an estimated 33 percent of soil is so degraded that its ability to grow crops is compromised. Meanwhile, monoculture—the strategy of sowing huge fields with a single crop—achieves higher yields but also puts more pressure on soil and increases the risk that plants will succumb to pests or disease.

Many of humanity's solutions to these problems also create other issues, Land Institute researchers say. Fertilizer can counter soil degradation, but it pollutes waterways and produces nitrous oxide, a potent greenhouse gas. Pesticides might reduce threats from insects, but they destroy other vital species. Cover crops will curb erosion, but they can be difficult to plant and maintain.

And modern farming is hugely carbon intensive. Factoring in fuel for machinery and food transport, methane produced by belching livestock, and the carbon that's lost when ecosystems are converted to cropland, agriculture accounts for about a quarter of humanity's annual planet-warming emissions.

Yet farms are also threatened by climate change, which will increase the risk of prolonged droughts and catastrophic floods.

In Kansas, one of the nation's leading producers of wheat, these problems are on full display. The state loses an estimated 190 million tons of its rich topsoil each year. Climate change has made Kansas summers hotter and drier, but also makes rainstorms more intense. The state's farmers are among those most at risk of losing crops as a consequence of human-caused warming.

"It's a disaster," Tim Crews, the Land Institute's lead soil ecologist, tells me one damp day in June. Our shoes squelch in the mud as he leads me around the institute's Salina, Kansas, campus. As we talk, the rain is almost certainly destabilizing soil and washing it into surrounding streams.

Crews sweeps his hand out, as if to indicate not only the farm fields across the road but the entire U.S. agricultural system.

"This is the ecosystem that feeds us, and it has just been nuked," Crews says. "Is this really the best we can do?"

The Seed

Land Institute scientists disagree about how to describe what they're proposing. Is it a natural evolution from the past ten thousand years of annual agriculture? Or something more like a midcourse correction?

Rachel Stroer, the Land Institute's president, calls it a "paradigm shift."

"Instead of an annual monoculture," she says, "we're trying to create a perennial polyculture"—cultivating diverse mixes of long-lived plants.

"We want to create an agricultural system to feed humanity that uses nature as the measure of success."

Before people started intensively farming here, Kansas boasted some of the richest soils on Earth. In native prairies, dozens of grass species intermingled with clover, wildflowers, lichens, and shrubs, their roots extending as far as fifteen feet into the ground. Periodic fires sparked by lightning or set by Native people helped clear debris and promote new growth. Insects, birds, prairie dogs, and buffalo foraged in the vegetation, while millions of munching microbes buried carbon and other nutrients deep in the earth.

"The ecosystems that built the soils upon which we eat today, and that we have degraded, were perennial and diverse," Stroer says. "That's where we get those two characteristics that we're trying to bring back into agriculture."

Yet proponents of perennial polyculture have a problem: more than half of all calories consumed by people come from grains, and no one has ever domesticated a grain that lived beyond a year.

That challenge falls to plant biologists such as Lee DeHaan. The son of a Minnesota corn and soy grower, he'd heard family members talk about the Land Institute's ideas with some skepticism.

"But it captivated me," he says. "I saw it as solving food for humans, environmental problems, and financial security for farmers."

He began experimenting with a wild grain known as *Thinopyrum intermedium,* or intermediate wheatgrass. Originally from the steppes of Europe and Asia, it had been brought to North America as forage for cattle, but scientists had a hunch it could also feed people.

In the early 2000s, Land Institute scientists planted their first plots of intermediate wheatgrass. When the plants matured, DeHaan and his colleagues selected the one thousand top specimens to replant. And when *those* plants matured, they chose the best among them for further breeding. It was the same process that farmers have been using to domesticate crops for millennia.

To the scientists' surprise, those early harvests were wildly successful. The new batch of plants had stronger stalks and bigger seeds that didn't fall out of their husks before they could be harvested.

"We started to realize we were not that far away from something farmers could actually use," DeHaan says.

"But the original domestication of crops took hundreds and thousands of years," he adds. "And with climate change, we don't have that much time."

So, he turned to tools that were unavailable to his ancient predecessors: gene sequencing, artificial intelligence, and advanced supercomputers. Once DeHaan identified the genetic markers associated with the traits he was looking for, he didn't need to wait for the plants to fully mature before picking the best ones to breed.

After two decades and eleven cycles of this process, the Land Institute has domesticated a form of wheatgrass whose seeds are two to three times bigger than those of its wild ancestor. Under ideal conditions, it can provide as much as 30 percent of the yield of traditional wheat. They call their trademarked creation Kernza—an amalgamation of "kernel" and "Kansas."

But the plant's best qualities are belowground. DeHaan shows me a photograph of Kernza's roots hanging in a Land Institute stairwell—the life-size image is so long, it takes up two stories. In the first four years after planting, Land Institute research suggests, a one-acre plot of Kernza will pull roughly 6.5 tons of carbon dioxide out of the air and into those deep roots.

Kernza can't completely replace regular wheat—at least, not yet. As Jenny kneads our bread dough, she explains that the weaker gluten proteins in Kernza flour make it harder for loaves to hold their shape. And because Kernza grains are so small, the flour also has proportionally more bran, the hard outer coating of a grain. This isn't necessarily a bad thing—bran is full of fiber, protein, and other nutrients. But it's not exactly ideal for making angel food cake.

Still, mixed with an equal amount of whole-wheat bread flour, it's shaping up to make a good-looking loaf. Jenny places the dough inside a cast-iron cooking pot, which will help the bread bake evenly, and slides it into the hot, waiting oven.

The Harvest

Before those grains arrived in my kitchen, they were grown by someone like Brandon Kaufman, a fourth-generation Kansas farmer. Kaufman is one of the cofounders of Sustain-a-Grain, a coalition of growers and buyers working to turn the Land Institute's vision of perennial polyculture into a marketable reality.

That means more than just planting Kernza. Farmers must also figure out how to cultivate it alongside other species, creating fields that are diverse as well as deep-rooted.

I visit Kaufman on a sparkling summer morning, driving past endless rows of corn, soy, and wheat that blanket central Kansas. The orderly fields belie the tumult facing many small farmers. Net cash income for farms in McPherson County, where Kaufman lives, fell by half between 2012 and 2017, according to the U.S. Department of Agriculture. Buying seeds, fertilizer, and equipment can put farmers in the red before a single grain is harvested, and natural disasters—which are growing worse because of climate change—can wipe out a whole year's work in a single day. The combined debt of all U.S. farmers totals more than $400 billion.

Compared with more-traditional farms, Kaufman's plots look somewhat scruffy. Tufts of chicory, alfalfa, and clover are interspersed with the tall stands of Kernza. Ladybugs dot the greenery, and songbirds twitter in the brush. Kaufman leans down to turn over a dried clump of dung—an offering from the cattle he brings to graze here twice a year. Wriggling in the exposed dirt are several soil-enriching earthworms.

Kaufman's neighbors would call his fields "dirty." The mix of crops makes them harder to harvest by machine and less profitable per square foot. His own father, who gave him this land, is skeptical of the whole experiment.

Yet Kaufman says perennial polyculture has been profitable for him. He points out the rich, dark green color of Kernza growing beside patches of alfalfa—a product of the latter plant's ability to fix nitrogen in the soil. When he brings his cattle to eat the alfalfa, they will spread their waste across the fields and trammel old vegetation into the earth. All this means Kaufman doesn't have to buy synthetic fertilizers or spend time hauling manure. The ladybugs and birds feed on crop pests, reducing the need for pesticides.

"I don't need all these inputs and overhead," Kaufman says. "Diversity is my crop insurance."

That's not to say it's easy. Kaufman is in a constant battle with weeds, which flourish in his herbicide-free fields. Farm equipment isn't designed to handle Kernza's small grains, so harvesting and processing are less efficient. There are scores of kinks to work out in the supply chain connecting farmers to consumers.

But Kaufman thinks about the land he inherited, depleted by a century of intensive farming. He thinks about the state of the planet, battered by climate change and species loss and habitat destruction.

And he thinks about his four children, who he hopes will someday earn their livings from this earth. If his experiments with Kernza are successful, he'll be able to leave them not just a healthier farm but a healthier world.

"Talk about a legacy," he says.

The Meal

Two decades after the Land Institute planted its first field of intermediate wheatgrass, Kernza can be found in the ingredient lists of cereals, baked goods, and beers. For now, most of the products are pricey—the flour that Jenny and I are baking with costs more than eleven dollars per pound, for example, compared with less than a dollar per pound for regular all-purpose flour.

Meanwhile, DeHaan and colleagues around the world are working on perennializing other crops: soybeans, sorghum, sunflowers for oil. A form of perennial rice developed at Yunnan University in China has been in commercial production since 2018.

"There's a lot more belief we can achieve what once seemed unachievable," DeHaan said.

The proof will be in the eating. Jenny pulls our loaf from the oven, filling the kitchen with a tantalizing, yeasty smell.

"I'm excited that there's movement in the idea of more sustainable agriculture," she says. "I hope this can prove there's a market."

Finally, the bread is cool enough to cut into. Jenny takes a bite, tilts her head, and chews. "It tastes like . . ." she trails off, then tries again.

"Texturally, it's like rye, but a little spongier," she says. "And it's almost like it's got a hint of herby or spicy-ness."

She grins. "It's delicious."

And we both grab another slice.

Power Shift

FROM The Verge

THE SUN WAS STILL SHINING when Marcia McWilliams lost power in her New Orleans East home. She had been cooking steaks on her stovetop on Sunday, August 29, just before Hurricane Ida would tear through the city, and was planning to hunker down in her two-story house with her husband, elderly uncle, and granddaughter. But the blackout had arrived early. Up and down the block, neighbors ventured outside to check in with one another. "We're all looking at each other like, 'What's going on?' The sun was shining!" McWilliams says.

An aging network of power lines connects McWilliams's home and her neighbors' to energy sources outside of the city. Ida's hurricane-force winds started severing those connections soon after the storm made landfall in Port Fourchon, about sixty miles south of New Orleans. When the storm reached McWilliams's home, it brought damage to accompany the darkness. Part of the ceiling—weighed down by water—collapsed on her husband as he tried to put out bins beneath the leaks.

When skies were clear again, the blackout on McWilliams's block lingered. McWilliams hung out on her front porch where she could catch a breeze despite the summer heat and listened to local news on her uncle's battery-powered radio. "It was just total chaos. We're listening to everything, and I'm just getting angrier and angrier," she says. "That was not supposed to happen. We should not have lost electricity when we just built this power plant."

"This power plant" was the brand-new gas plant built less than six miles away from McWilliams's house. She had fought against

the plant's construction for years prior, worried about higher utility bills and pollution that it would bring to the neighborhood.
But in Ida's wake, she expected the plant to fulfill a promise that
had convinced New Orleans's previous city council to approve its
construction in the first place: that it could quickly restore electric
service if a major storm ever wiped out power across the entire city.

The New Orleans Power Station is what's known as a peaker
plant, designed as a backup source of electricity whenever there's
a power shortage. The city gets nearly all of its electricity from
gas, coal, and nuclear power plants outside of its borders. If a
disastrous storm cuts New Orleans off from those outside power
sources, the peaker plant could provide some locally generated
electricity to tide residents over.

Crucially, the New Orleans Power Station was designed to be
able to perform a "black start"—meaning it can start up on its own
without needing a jolt from the grid. Ida had devastated the electricity grid, taking out all eight transmission lines that bring power
into the city, creating exactly the type of scenario for the gas-fired
plant to prove that it was worth its $210 million price tag.

It never performed a black start. Residents sat for days without
power during scorching heat and stifling humidity. Widespread
outages lingered across New Orleans until September 9, eleven
days after Ida. Without power, most people couldn't find relief with
air-conditioning, fans, or ice. More people died in New Orleans
from the heat that followed Ida than perished during the height of
the storm's wrath. Outside of the city, residents in more rural areas
were left in the dark for more than a week longer.

Now, New Orleans faces tough decisions ahead as Louisiana
continues to recover after Ida. What should the wrecked grid look
like in the future? Can it be safeguarded from another storm?
Who gets to call the shots?

For McWilliams and some of her neighbors who came together
to try to stop the peaker plant from being built, the answers were
clear long before Ida hit. They want a shift in power: from dirty
fossil fuels to clean energy and from top-down decision-making to
bottom-up solutions.

Louisiana has long prided itself as an energy state, although that's
historically been oil and gas. The colors of New Orleans's beloved
Saints football team are black and gold after oil, "black gold," that

made the team's first owner rich. New Orleans East similarly got
its moniker from oil-rich tycoons based in Dallas who founded the
development firm New Orleans East, Inc., Sarah M. Broom writes
in her memoir, *The Yellow House*. In the late 1950s, the firm set out
to tame more than thirty thousand acres of swampy, mostly unde-
veloped land and turn it into a suburban arm of the city designed
primarily for white, affluent residents.

By the 1970s, refugees arrived from Vietnam and found a new
home in New Orleans East. A Catholic charity helped people escaping war resettle in low-income, subsidized apartments. Refugees coming from fishing villages in Vietnam found work in the
region's commercial fishing industry. They planted gardens ripe
with vegetables and herbs common in Southeast Asia: taro, bitter
melon, lemongrass, and more. The community blossomed to become one of the most concentrated populations of Vietnamese
people outside of Vietnam.

Green space and bigger homes attracted families like Beverly
Wright's when she was a young girl. "Christmas time came, all of
my relatives from in the city who lived in shotgun doubles on rag-
gedy streets saw coming to our house as a place of a reprieve,"
says Wright, founder and executive director of the nonprofit Deep
South Center for Environmental Justice. "We thought it was the
most beautiful place in the world."

The area remained a stronghold for middle-class Black families
and the Vietnamese American community in the city even after
an oil bust in the 1980s and subsequent white flight to nearby
suburbs. Its diverse neighborhoods span an area that now makes
up the largest part of New Orleans and still contains much of the
city's natural wetlands.

Dawn Hebert bought her home, a stately two-story brick house
framed by four pillars out front, across the street from Wright
in the 1990s. But the neighborhood hasn't really been the same
since levees failed to hold back Hurricane Katrina's floodwaters in
2005. Driving through New Orleans East, Hebert can rattle off the
names of businesses lining the road that never came back after the
storm. Off the I-10, there's the dusty abandoned lot where Lake
Forest Plaza used to stand, once the biggest shopping center in
the state. What was once a Six Flags amusement park is now over-
grown, empty ruins.

"I have been a big advocate, honestly, since Katrina," says He-

bert, who is now president of the East New Orleans Neighborhood
Advisory Commission. "Never in my wildest dreams did I think I
would be in the position I'm in right now," Hebert says of her work
holding the city accountable for how it treats New Orleans East.
She and other residents have grown used to having to stand up for
themselves.

Adding insult to injury, after the floodwaters receded, city officials proposed a plan to rebuild the city without including much
of New Orleans East—instead, shrinking the city's footprint and allowing wetlands to reclaim areas deemed too flood-prone to revive.

Public outrage ultimately kept New Orleans East on the map.
But the community soon became a sort of sacrifice zone for the rest
of the city's recovery. The remains of thousands of destroyed homes
and buildings needed to be buried. The city picked New Orleans
East to be the final resting place. A freshly dug landfill—which
lacked a clay liner to prevent nasty leaks—took in about 150,000
tons of garbage in four months before residents in the nearby Vietnamese American community successfully rallied to shut it down.

About a decade later, many of the same residents found themselves defending their neighborhoods yet again from another new
source of pollution: the gas-burning New Orleans Power Station,
to be operated by the local energy utility Entergy. It would be built
at the site of retired generators that were even more polluting.
"They try to put everything bad out here for our community that is
environmentally unsafe," Hebert says. After getting calls about the
power plant proposal from her neighbor Wright and other advocates, Hebert set up a meeting between the Deep South Center for
Environmental Justice, the East New Orleans Neighborhood Advisory Commission, and other groups to let the community know
about the plans.

It was the beginning of yet another standoff over the future of
New Orleans East. "Oh my God, that was a whirlwind," Hebert
recalls. At city council meetings, Hebert and her friends faced
apparent supporters of the power plant who showed up wearing
matching orange shirts and carrying signs with phrases like, WE
NEED POWER IN THE CITY. Some were actually actors, hired to
promote the plant. One of Entergy's subcontractors had paid
them between $60 and $200 each, the Lens reported, and the city
ultimately fined Entergy $5 million for the fiasco. Now, attendees
who want to give public comment at city council meetings have to

check a box at the bottom of a speaker card to identify if they are "a paid representative or receiving any type of compensation or thing of value in exhange [sic] for speaking or attending today."

Hebert and McWilliams had a crew of their own standing up for New Orleans East. "You had environmental injustice, you know, building these plants around African Americans, Vietnamese Americans," says McWilliams. "We were out there fighting the fight, fighting for them, fighting for us. We don't want this; we don't need this. Go with more renewable energy: wind and solar!"

Renewable energy would carry a few benefits for the area: it would improve air quality and reduce greenhouse gas emissions that fuel stronger storms. On top of that, critics of the plant believed there were better ways to strengthen the existing power grid than building a new fossil-fueled power plant. Aging power lines, which Entergy was previously fined for poorly maintaining, are an especially easy target for storms. So, if the city really wanted to harden the grid, it ought to upgrade and fortify those lines, experts said. But the way utility regulation works in the United States generally incentivizes companies to spend more money on new capital investments in things like new power plants rather than on maintenance costs for upgrading old power lines.

"Basically, we did not need the plant. And why are you paying for something that we don't need? It's just crazy," says Hebert.

This time, she and other New Orleans East advocates lost the battle. Entergy successfully sold the plant as a way to make the grid more dependable in times of crisis. "It will provide a reliable local source of power generation in Orleans Parish to help stabilize the grid and keep the lights on," then-president and CEO of Entergy New Orleans Charles Rice wrote in an opinion arguing for the new plant on Nola.com in 2017.

City council approved the plant in 2018, and it came online last year.

Sitting in the second row of city council chambers on September 22, Entergy New Orleans CEO Deanna Rodriguez spoke softly with colleagues. "I'm scared," she told one as she sat in the audience waiting for her turn to come forward before the city council.

Had she seen the latest investigation by ProPublica, the colleague asked her. "Is it on us? Propaganda?" Rodriguez responded, before telling him she didn't want to know more about it until after taking the stand. The story, written in partnership with NPR, had published that morning with the headline "Entergy Resisted Upgrading New Orleans' Power Grid. When Ida Hit, Residents Paid the Price."

The storm took out every transmission line the city relies on to bring in electricity. During the ensuing power outages, temperatures soared into the nineties while lingering humidity made things even more stifling. With nearly all of Southeast Louisiana without power, many residents couldn't find air-conditioning. The Orleans Parish coroner's office attributed nine of the fourteen casualties tied to Ida to "excessive heat during an extended power outage." That included seventy-three-year-old Iley Joseph, who was found dead in his sweltering home four days after the storm's landfall. Sixty-five-year-old Laura Bergerol was similarly found dead inside her home by a neighbor a full seven days after the storm had passed.

"Please stop acting like you're the victim. You are the goliath. You are a powerful Fortune 500 company with all the resources in the world, with record profits last year of $1.4 billion," councilwoman Helena Moreno said to Entergy officials as she opened up a meeting of the city council's Utility, Cable, Telecommunications, and Technology Committee.

Most utilities are regulated at the state level; New Orleans is the only city in the United States where the city council is responsible for regulating an investor-owned energy utility while there is already another energy regulator at the state level. The council oversees Entergy New Orleans, a subsidiary of Entergy Corporation, which provides power across Arkansas, Louisiana, Mississippi, and Texas. It gives the council unique power to be responsive to residents' needs, although the council has been criticized in the past for being ill-equipped to regulate Entergy. With elections coming up, members seeking reelection would soon have to answer to voters.

That day, the city council would vote on measures that would launch investigations into the outages, conduct a management audit of Entergy, and commission a study on potentially ending Entergy's monopoly in the city's energy market. Entergy had been pressuring them to let up.

The day before the meeting, Entergy New Orleans released a statement with suggestions for the city's energy future if it no longer wanted the subsidiary to provide everyone's power: Entergy

New Orleans could merge with Entergy Louisiana, sell the utility, or spin off the business to create a stand-alone company. Or the city could create a municipally run utility instead. It was a move that some people in the energy business described as a dare or a temper tantrum in response to city council's scrutiny. Councilwoman Moreno called it a "ploy" for the utility to slip through city council's fingers and find a friendlier regulator.

"If you hear my voice shake just a little bit, it's because I'm nervous," Rodriguez said before she started her testimony before city council. Many of the questions she faced drilled down to one thing: What exactly did the $210 million power station actually do after Ida? Residents were paying for the cost of the new plant with higher utility bills, even if it didn't do what they thought it would do in a catastrophe.

Rather than perform a black start at the power station, the quicker and safer way to restore power was to get one of the transmission lines back online, Rodriguez said. With a jolt of power from the transmission line, the plant then fired up and helped distribute some electricity throughout the city.

The power plant hadn't prevented extended blackouts, though. Some people—including Dawn Hebert—got power back within two days, while others waited ten or more days. Even after firing up, the plant wasn't able to get electricity to places where distribution lines were still down. The power station wasn't designed to generate enough power for the entire city. Instead, it had capacity to meet just under 12 percent of the city's needs, and the power first flowed to critical infrastructure like hospitals after Ida passed through.

"New Orleans Power Station did what it was designed to do during and following the storm," Entergy said to the Verge in an email. The utility company referred to residents' expectation that the plant would prevent outages as "an unfortunate misunderstanding influenced by inaccurate reporting." Power plants are designed to shut down when lines connecting them to customers are damaged or destroyed, it said.

Even as power started to come back on, Entergy's outage maps had glitches. They showed that power was restored in McWilliams's area even though it wasn't—seeing the alerts, some residents planned to check out of hotels where they had taken shelter and return. McWilliams spent the days afterward fielding calls from neighbors who had evacuated, telling them it still wasn't safe to come back. "No, we do not have electricity, do not come back yet," McWilliams told them. She didn't get electricity back in her home until a week after the storm.

Rodriguez carefully answered questions from council members, prefacing by saying that a class action lawsuit filed against the utility by customers limited how much she could share. When city council finished asking questions, it was time for public comment. As the first speaker took to the podium, Rodriguez and other representatives from Entergy got up from their hot seats in the front of the chambers and walked out of the room.

The majority of comments were made online, which city council began accepting during the COVID-19 pandemic. There were too many to get through, so a designated speaker read as many as they could in twenty minutes. One woman described riding out the outages with her four-year-old child. "I tried my best to be cheerful and positive for my first real experience with a hurricane, being eight and a half months pregnant made it nearly impossible to maintain good cheer in the unmitigated late summer heat," she shared.

When it was her turn, Hebert stepped up to the podium in a sky-blue blouse over white pants and addressed the elephant that was not in the room. "I got to say, the fact that all of them [from Entergy] left means I don't think they care too much," Hebert said. "Entergy New Orleans should be held accountable, and changes that should be made are in your control," she said to the council members.

City council members, responding to Hebert and other speakers, decided to call Entergy representatives back into the room. They returned about forty-five minutes after leaving, claiming to have been watching from outside.

In New Orleans East, just seven miles of backroads and highway separate the post-Katrina landfill, gas plant, and the potential beginnings of a clean energy future. Several months after the gas peaker plant came online in New Orleans, Entergy opened a new solar farm just down the street. It's at NASA's Michoud Assembly Facility, which the agency calls "America's Rocket Factory."

Also nearby is the nation's only private testing facility for cutting-edge offshore wind turbine blade technology, according to the regional economic development organization Greater New Orleans, Inc. The Biden administration recently moved to open up the Gulf of Mexico to offshore wind, which some Louisiana business leaders and lawmakers see as a big moneymaker for the state in the future.

Policy makers have begun to accept a difficult truth. Greenhouse gas emissions from fossil fuels are supercharging the weather. Hurricanes pack bigger punches than they did in the past. Heat waves are becoming more unbearable, too. And rising sea levels are eating up the state's coastlines. Together, these disasters will deal repeated, worsening blows to residents and the electricity grid. The only way to stop or even slow that trend is to tackle climate change, which requires slashing the pollution that comes from coal, oil, and gas. And if Louisiana wants to stay an "energy state," it will have to find cleaner sources.

Governor John Bel Edwards signed an executive order last year setting a goal for the state to cut or cancel out all of its planet-heating greenhouse gas emissions by 2050. In May, the New Orleans City Council passed a new energy standard that pushes Entergy to get all its electricity from "clean" sources by the same date. Entergy says it wants to increase its reliance on renewable energy to meet a company goal of reaching net zero carbon dioxide emissions by 2050—keeping it in line with the state's targets.

It might sound like a short timeline for an energy state to kick its fossil fuel habit, but the deadline falls in line with what research has found is necessary globally to keep temperatures from rising to a point at which human civilization would struggle to adapt. The Biden administration is similarly trying to set the United States on course to reach net zero greenhouse gas emissions by 2050. On the way there, Biden envisions a "100 percent" clean power grid by 2035. So, what plays out in New Orleans is a microcosm for the energy transition that's ahead for other oil and gas towns and the United States as a whole.

Slowing down climate change is a global problem. But adapting to changes that are already here—like stronger storms—will require local solutions. That's especially true for electricity grids like New Orleans's, prone to climate-fueled disasters. Nearly all of the city's electricity comes from somewhere else, some of it sourced from generators as far away as Mississippi and Arkansas. That's typical of how electricity grids work today. Giant fossil-fueled power plants and nuclear reactors generate enough power for expansive areas. Electricity from the plants zips along transmission lines, which transport power across long distances at high voltages. The electricity makes a pit stop at a substation near its final destination, sometimes hundreds of miles from the power plant, where the voltage is tamped down so that it's safer to move through neighborhoods via local distribution lines.

Each of those points—power plant, transmission line, substation, distribution line—could be vulnerable to a disaster like Ida. A storm only needs to take out one link in the chain to cut off power to many communities.

Shrinking down the grid and generating power locally is one way to minimize those vulnerabilities. In fact, there's already a blueprint for how that could work. In 2017, the Department of Energy's Sandia National Laboratories and Los Alamos National Laboratories worked with the city to come up with a plan for how it could make the energy system more resilient to future hurricanes. That research identified advanced microgrids as one potential solution.

"We cannot be thinking about the electric system in that linear way anymore," says Logan Atkinson Burke, executive director of the local consumer advocacy group Alliance for Affordable Energy. "Think about the whole of the system, rather than thinking about the way the system was built 120 years ago."

Here's how advanced microgrids might work: First of all, energy is generated locally. That can bypass one potential weak point— transmission lines that bring power in from faraway places. Solar panels, in particular, are easy to spread around the city and still tie together so that they can work in tandem. They could be clustered together in a solar farm, like the one at the NASA facility. Or they could be spread across homes and buildings throughout the city. More than 90 percent of rooftops in the city of New Orleans are well suited for solar generation, according to one Google analysis.

There was already a boom in residential solar after Katrina, thanks to post-storm tax incentives. The city ranked fourteenth in the nation for the most solar capacity installed per capita in 2020, ahead of sunny Los Angeles. But to be helpful in a crisis, panels will also need to be paired with batteries that can store extra juice for a rainy day. And local distribution lines will also need to be sturdy enough to keep all those solar panels and batteries connected to one another.

If disaster strikes, especially if it knocks out other parts of the larger grid, an advanced microgrid can automatically cut itself off and act as a sort of energy island. Outfitted with an advanced microgrid, New Orleans East would be one area that's particularly well suited to act as a "resilience node," according to the Sandia analysis. That's where a relatively small amount of backup generation could power a high concentration of critical infrastructure serving a large population.

"We've been urging Entergy to look at things like microgrids," Burke says. "Instead, all the money that Entergy has been spending is on their gas infrastructure."

Entergy tells the Verge that it spent $3.9 billion on its transmission and distribution systems between 2016 and 2020. "It is important to invest in generation, transmission, and distribution, and Entergy has no financial incentive to prefer one over another," Entergy said to the Verge in an email. "Physics dictates how we invest to keep the system reliable and keep the lights on."

On an even smaller scale, solar power and batteries can benefit individual homes. That's how Devin De Wulf's household kept the lights on during and after Ida. His family weathered the blackout with his refrigerator, ice machine, and fans. He avoided using the central air-conditioning because that would have taken up too much power, but he's since bought a small window AC unit that he can plug in and power on with the batteries the next time a storm knocks power out.

De Wulf, who founded nonprofit organizations that bring food to essential workers, musicians, and artists during the pandemic, helped out his neighbors with his solar system. He set out a power strip on his porch so that others could charge their phones or other devices. And he ran an extension cord to a neighbor two doors down so that he could run his oxygen machine.

"Nobody cares about us, but us," DeWulf tells the Verge. "We have to be the first responders."

DeWulf's next project is an initiative to get solar and batteries to restaurants around the city so that they can do the same thing for their communities.

In the absence of a disaster, microgrids can still function as part of the larger grid system that connects the rest of Louisiana and nearby states. When the larger grid is up and running, a household can sell its excess solar power back to the utility to lower its bills.

That bidirectional flow of power is crucial for any future grid that's built on more renewable energy. It's another reason why Burke and other advocates have called for more investment in upgraded transmission lines. Sunshine and wind gusts are intermittent sources of energy—they're renewable, but they can only be tapped at certain times. Luckily, they can also fill in for each other. When the sun is shining brightly on the Gulf Coast, the region can send its solar energy up to the colder Midwest. Midwestern states and Texas, where a lot of wind energy is generated, can, in turn, send their renewable energy down in exchange.

Instead of the linear grid system where electricity flows in one direction from polluting generators, it's a system built on two-way communication. And even though utility scale wind and solar farms (paired with batteries and other power sources like hydroelectricity) will be needed to meet everyone's electricity needs, a single home solar system can still contribute when its power is needed.

Grids in other climate-stressed places are already evolving with the help of renewable energy. In California, where more intense wildfire seasons have sparked blackouts, solar panels and batteries spread across apartment buildings work together as a virtual power plant. There are similar projects in Utah and Arizona. Even the batteries in electric vehicles could one day be tied together to provide backup power in disaster scenarios.

Dawn Hebert left New Orleans before Hurricane Ida hit, sixteen years to the day Hurricane Katrina made landfall. If her home got as much damage from Ida as it did during Katrina, she wasn't going back, she told herself. Hebert had evacuated then, too, and came back to find that her home had flooded with six feet of water. "The furniture must have just been flowing from room to room," she says. Even the refrigerator and heavy china cabinet were knocked over.

"That's why I don't even have a living room setting anymore," she says. "Because it's like—what's the point?" She has two plush chairs facing each other next to a large window looking out onto the front yard. There's a crib and play mat for a grandson she babysits during the day now that she's retired from working at the post office. But the room is relatively sparse when it comes to furniture.

Thankfully, after coming home to New Orleans East after riding out the storm in Mississippi, Hebert saw little damage to her house in comparison to the destruction during Katrina. While levees failed catastrophically during Katrina, they held up this time. It was a testament to how preparing infrastructure for a more brutal future can prevent a lot of pain.

There's still pain that's tangible in the community. "New Orleans East is this place that has this very palpable complex trauma that's alive because of the environmental racism," says Jacqueline Thanh, executive director of VAYLA-NO, a nonprofit born out of a Vietnamese American community-led campaign to stop the landfill post-Katrina. Residents have seen their homes flooded, nearly wiped off New Orleans's map, saddled with garbage and pollution, and then left in the dark after Ida. For refugees, all of that came after being pushed from their homes in Vietnam by war.

"Looking at New Orleans, I worry about its survival for people of color," Wright tells the Verge a couple weeks after Ida. "I don't worry about its survival for rich white people. This will always be their playground. But I worry about the Indigenous population and people of color and whether or not we could weather the storm." Just before Ida hit, Wright had finally been able to finish repairing damage Hurricane Katrina had inflicted on her mother's house. Luckily, the home was mostly spared from Ida's wrath.

Whenever top-down systems and infrastructure failed them, New Orleans East residents found ways to persist. There's certainly talk, especially among younger residents, of picking up and leaving. But community advocates see a future worth fighting for. "Imagine the future for New Orleans East where young people can thrive—not just survive it and leave it," Thanh says.

Structures that have failed in the past can be strengthened so that it isn't vulnerable communities' resilience that's so harshly tested when the next disaster hits. Instead of a gas-fired power plant in their neighborhood, McWilliams and other advocates want to see more solar panels. And after decades of being treated as an afterthought or a dumping ground, they see New Orleans East leading the way to a more sustainable future.

As the days lingered on without power after Ida, McWilliams coped with the heat with the help of an ice pack that her granddaughter, a nurse, was eventually able to bring to her. "Nana, we just need to move away," McWilliams recalls her granddaughter once saying to her. "I said, 'Well, baby, it's not as simple as that. You can't run from Mother Nature.'"

LUCY SHERRIFF

Beavers Are Firefighters Who Work for Free

FROM *Sierra*

SINCE KENNETH MCDARMENT was a kid in the 1980s, he's seen the foothills of the Sierra Nevada change. As a councilman of the Tule River Tribe, a sovereign nation of around a thousand members living on 56,000-odd acres in the foothills of the Sierras, McDarment deals with everything water-related on the reservation. Today there's less rain and less snow than there was even a decade ago, which means that the land in the foothills was dangerously dry during the last fire season, when wildfires were sweeping across the state. "If you don't got water," says McDarment, "we don't got nothing."

So, in 2014, McDarment began looking into getting ahold of some beavers. McDarment hoped that beaver dams would create soggy areas on tribal lands that wouldn't dry out during heat waves. "We're hoping that means our land will be less likely to burn during fire season," he says. "Beavers were here originally. So why not bring them back and let them do the work they do naturally?"

There was just one problem—it is illegal to move beavers without a permit. And a permit to move a beaver isn't easy to come by.

Killing Is Easier Than Paperwork

If a farmer, landowner, or property developer wants to get a beaver out of a certain area, it's easier to kill the beaver than to apply to move it elsewhere. Across the states, it's common for landowners to dynamite beaver dams, with whole forums dedicated to the topic and dramatic instructional YouTube videos.

In 2019, the California Fish and Wildlife Department issued 187 depredation permits to kill beavers across the state. In 2020, that number rose to 204. While not all permits are necessarily fulfilled, it's also true that multiple beavers in a single area can be killed under one permit. Despite the fact that beavers once roamed far and wide across the state's waterways, it's illegal under California law to release one into a new location. Though beavers are native to the state, they weren't recognized as such by California Fish and Wildlife until 2013.

The beaver does more to shape its environment than nearly any other animal on Earth. Beavers can cause incredible amounts of destruction to infrastructure, downing power lines and blocking and rerouting waterways. But their dam-building also can improve water quality, reduce flood risk, and create the conditions for complex wetland habitats to form—providing refuge for wildlife and storing carbon in the process.

"It's not that complicated," says Joe Wheaton, an associate professor at Utah State's Department of Watershed Sciences, who developed the university's BRAT project (short for Beaver Restoration Assessment Tool). The initiative serves as a planning aid for researchers and restoration managers who are looking to assess the potential of beavers to restore watersheds. Wheaton has worked on the Tule River Tribe's reintroduction project and many others across the states. "If you wet up the sponge of your valley bottom, you have the potential to at least slow the spread, if not at least have the land act as a livestock and wildlife refuge during wildfires. If you have a wide enough valley bottom, and beaver are present, it can be big enough to actually stop the advance of these wildfires. That information just needs to get out there."

If You Can't Catch a Beaver, Build It a House

The Tule Tribe wasn't the only one trying to bring back beavers to manage the land. The Tulalip Tribes, situated in the western corner of Washington, noticed around 2012 that beavers had started to make a comeback, but only in the lowlands—beavers prefer lower elevations, where the water flow is slower.

Since Washington also banned beaver relocation, the tribe began building beaver analogs—essentially man-made beaver dams—to entice the animals into returning. And it worked. The beavers began using the analogs as footholds to build their own dams. In this case, says wildlife biologist Molly Alves, who led the tribe's re-introduction, the goal was to create better conditions for salmon. Beavers not only carve out channels and streams that leave behind cooler water—and improved water quality—for salmon, but their ponds also create ideal nurseries for juvenile fish. "We wanted to return beavers to their historic habitat," Alves says. "It was from a salmon habitat restoration perspective."

After months of to-ing and fro-ing, the Tulalip were able to leverage their sovereign rights as a tribe to get a letter of excep-tion from the Washington State Fish and Wildlife Service, which allowed them to relocate beavers to their lands beginning in 2014.

The program was a success. The beavers adapted to their new habitats and quickly began to create better salmon habitat. Within a few years, Alves and the team of scientists and tribe members working on the project began receiving calls from nonprofits, county officials, and landowners from all over the state, asking how to move beavers without having to kill them. Word of the tribe's work had spread around the state.

Because the Tulalip were operating under sovereign rights and had a letter of exception, they were the only nonlethal beaver removal entity in the whole of western Washington. Callers com-plained about beavers on their property, flooding roads, and felling trees. "They were frustrated and angry because they didn't under-stand how to deal with the problem," says Alves. "They wanted the beavers to stay alive, but they didn't have any means to do that."

Tackling the Beaver Bill

Alves and her team began working with the Tulalip's team of lob-byists to try to amend HB 1257 (also known as the Beaver Bill), which limited the release of wild beavers, to make it possible for non-tribal groups to relocate beavers to any preapproved private land as long as the habitat was deemed suitable and unoccupied via a GIS model and in-person assessment and Washington Fish and Wildlife gave its approval. Alves also worked with Washington

Fish and Wildlife on developing a Certified Beaver Relocator training program.

When Alves went to testify at the statehouse and senate in support of the amendment, legislators weren't sure what to make of it, she says. "People were giggling and were like 'Oh, the Beaver Bill.'" But in 2017, the bill was amended, and the Washington Department of Fish and Wildlife created a statewide pilot project, which is now in its second year.

One of the best parts of being involved, says Alves, is seeing the perception of beavers shift in real time. "Normally people were fine going out back with their shotgun and dealing with the problem themselves," she says. While plenty of landowners across the state still call wildlife control rather than the tribe if there is a beaver they want to get rid of, the Tulalip are also working with trappers who catch beavers on private land to donate those beavers to wetlands restoration projects.

It's also true, says Alves, that some landowners who have called the Tulalip to help them remove a beaver on their property have changed their minds after they learn more about how beaver management can aid watersheds. In areas where beavers have been relocated, the Tulalip found a 2°C cooling of water downstream of dams. The amount of surface water present on the landscape increased by 2.5 times. "We know we're going to have more, and more severe, fires," says Alves. "But we feel that with beavers, the land will have the best chance of surviving."

Dr. Emily Fairfax and the Case of the Missing Beaver Research

One thing that has been missing in the discussion of beavers and wildfires has been science connecting the two. But that is beginning to change. In 2018, Emily Fairfax, a young Ph.D. student studying hydrological science at the University of Colorado Boulder, saw a tweet posted by Joe Wheaton, of the wildfire-scorched landscape following Idaho's Sharps Fire, with a small patch of green at the center. "Why is there an impressive patch of green in the middle of 65,000 acres of charcoal? Turns out water doesn't burn. Thank you beaver!" wrote Wheaton.

Three years earlier, Fairfax had quit her job as a systems engineer and entered a Ph.D. program because of her obsession with beavers. She had a theory: if beavers were so beneficial to wetlands restoration projects and protection against drought, surely they must also protect against wildfires.

But she found herself struggling to find any previously published research on the subject. "It was no-man's-land," says Fairfax, who found plenty of research on beavers, fish, and waterways, but none on beavers and fire. "When you try to do new research, it really helps when you can stand on the work of previous scientists. After a certain amount of time, after a question hasn't been studied, you start to think, 'Oh, it's because there's nothing there.'"

Instead, her leads came through people like Wheaton, and an educational site called Beavers in Brush, which aggregates information about prescribed burns, as well as rewetting the lands through beaver protection. "That made me realize this has merit, there are people who are aware that this can work," says Fairfax. "I don't know why people haven't studied this, but obviously this is a thing."

Fairfax began to carry out the scientific research that she had hoped to find. Using satellite images, she mapped vegetation around beaver territories before, after, and during wildfires (footage of wildfires in progress can show how a fire moves through a landscape). She visited field sites in California, Colorado, Idaho, Oregon, and Wyoming and found sections of creek that did not have beavers were on average more than three times as affected by fire—burning a bigger area—than areas where beavers had built dams.

"I expected some of the time beaver dams would work," says Fairfax. Instead, she found the presence of beavers had significant effects. "It didn't matter if it was one pond or fifty-five ponds in a row. If there were beaver dams, the land was protected from fire. It was incredible."

Redamming the Golden State

Fairfax hopes her research will help change California's strict rules around beaver relocation, the way policy is already changing in Washington, especially as wildfires in California have reached record-breaking levels over the past several years. In 2017, while McDarment was still trying to get permission to relocate beavers to tribal lands, the Pier Fire consumed 8,800 acres of Tule River tribal lands, including several giant sequoias.

McDarment's plan is to have beavers living throughout the reservation's 56,000 acres. "I'd like to see them slowly evolve and move into every stream and creek we've got," he says. "I'm really excited to get started. I just can't wait until we get to that point where we're bringing a family of beavers here." A number of other tribes in California are also exploring how they can reintroduce beaver to tribal lands.

Meanwhile, Fairfax's research on beavers and wildfires is only beginning. "I set out to ask a question: Do beavers keep the land green during fires, yes or no?" she says. "The answer was yes. But that's not the end of the story. Why? How? Does this happen everywhere? What if you have a tight canyon? I'm digging into the specifics now, so people can implement this and actually use beavers for fire prevention. I would love to be able to call someone up and tell them how many beaver dams they need in their creek.

"Right now, I have so little advice on how to do it. But at least I can now say it works."

JANE C. HU

New Wind Projects Power Local Budgets in Wyoming

FROM *High Country News*

WHEN THE COVID-19 PANDEMIC HIT, Robin Lockman feared the worst for her town of Cheyenne, Wyoming. As the city's treasurer, she estimated that it might lose up to 25 percent of its budget as tax revenues stalled and the prices of oil, gas, and coal tanked, eliminating money the city typically receives from the state as royalties from the extractive energy industry.

So, the city did the hard work of laying off eighteen employees and cutting funds for travel and training. And then a surprising thing happened: The huge deficit never arrived. In fact, over the summer, the city brought in more tax revenue than the year before.

Between July and September, Cheyenne saw a 20.5 percent increase in tax revenue compared to 2019. In September alone, the increase was a staggering 83 percent, or $1.4 million. "I was in shock when I saw it," said Lockman. She feared the good news was a mistake, so she called the Wyoming Department of Revenue to confirm the numbers. "The tax reported was legitimate, and was due to the Roundhouse Wind Project," said Lockman, referring to an energy development west of the city.

Throughout Wyoming, counties typically depend on industries like coal, oil, and gas drilling, mining, or tourism and recreation to bring in the taxes necessary to pay for education, community programs, and infrastructure. Overall economic activity is down—statewide, sales and use taxes shrunk 6 percent over the last year—but revenues from mining, quarrying, and oil and gas extraction are down nearly twice that. Meanwhile, profits from wind energy developments, like the Roundhouse Wind Project, are booming. Now, residents and officials are asking whether wind energy can help the state survive the economic storm of the pandemic—and become a reliable revenue stream for the future as fossil fuel income dries up.

Over the last decade, investors have laid the groundwork for wind farms across Wyoming, which ranks among the nation's top ten states for wind capacity, according to the U.S. Department of Energy. The Power Company of Wyoming, a private company run by the Anschutz Corporation, is developing a facility on Carbon County's 320,000-acre Overland Trail Ranch, just outside Rawlins. Framed by I-80 to the north and Sage Creek Road to the west, the property is one of Wyoming's classic wide-open spaces: miles of rolling grassland dotted with scrubby sagebrush. On land long used for cattle and sheep ranching, in a county named for rich coal deposits, construction crews have built roads and turbine pads for the Chokecherry and Sierra Madre Wind Energy Project, which will comprise nearly a thousand wind turbines by 2026. Sixty miles east, Rocky Mountain Power, a subsidiary of the state's largest utility, has begun erecting 270-foot-tall turbines at the Ekola Flats Wind Energy Project.

The projects will create local jobs as workers operate and maintain the facilities. "Once the Chokecherry and Sierra Madre Wind Energy Project is complete, we estimate 114 permanent jobs will be created," said Kara Choquette, Power Company of Wyoming's communications director. The Two Rivers and Lucky Star wind projects, smaller developments near the border between Carbon and Albany counties, are expected to generate another twenty-four long-term operations and maintenance jobs.

Right now, that new construction is generating much-needed sales taxes on materials. Between April and June, Carbon County's taxable sales increased by 108 percent. "Everyone else around here's hair is on fire, but we're sitting in a pretty good position, thanks to wind energy," said Dave Throgmorton, director of the Carbon County Higher Education Center, which is funded by the county and provides college-level and vocational training classes.

But that increased tax flow is not guaranteed to continue. Lockman, Cheyenne's treasurer, thinks the boom her city saw in the third quarter of 2020 might be a onetime influx from a period of heavy construction. "Personally, I feel it's probably a temporary type of situation, but I'm hoping it will be longer term," she said.

Residents have also raised concerns that wind energy jobs aren't going to locals. Many wind projects have been completed by teams that travel to the state for a few weeks at a time, living in camps at the edge of town, Throgmorton said. "Everyone was hoping locals would be involved in wind turbines, but turns out that's all done by specialty teams from outside," he said. For example, the permit applications for the Two Rivers and Lucky Star wind projects explicitly mention "man camps" and estimate that roughly 80 percent of workers will be "non-local." About half of the hundred people working on the Chokecherry and Sierra Madre project between April and November were from Wyoming, said Choquette, the Power Company of Wyoming communications director.

The wind industry faces other sources of resistance in Wyoming. Groups across wind-rich counties have opposed projects, saying turbines disrupt dark sky areas, impede views, or decrease property values. In addition, legislators have proposed raising taxes on wind generation and removing tax incentives. While those changes could generate more tax revenue, they could also drive wind projects elsewhere. According to a 2019 analysis by Wyoming's Center for Energy Economics and Public Policy, Wyoming already taxes wind developments at roughly twice the rate of neighboring Colorado.

Though legislators have not made any recent changes to wind taxes, recurring proposals scare off companies, said Terry Weickum, a former Carbon County Commission chairman and a newly elected member of the Rawlins City Council. "No one wants to do business in an unstable environment," said Weickum. "It'd be like if you bought a car for a hundred dollars, and then you found out this hundred-dollar car needs a two-thousand-dollar motor."

Some see it as a sign that the state is giving up on its sunsetting oil, gas, and mining industries. Throgmorton said many of his students imagine following in their parents' footsteps and working at the Sinclair Oil Refinery. "But that's more fantasy than real now," he said. Still, many Wyomingites hope that renewable energies like wind will help sustain the state's energy economy in the long-term. "Honestly, I hated [wind energy] when it first came along, but I realized I didn't know anything about it," said Weickum. "People who have made their living in oil, gas, and coal—they feel like if you like wind, you're cheating on your wife, you're cheating on oil, gas, and coal. But we need every piece of it."

Even if the wind industry doesn't continue generating hefty tax revenues, its contributions could help keep counties in the black. Companies pay sales tax on equipment purchased for maintenance, like replacement turbine blades, property taxes on the assessed value of projects, and taxes on the energy generated. According to Connie Wilbert, director of the Sierra Club's Wyoming chapter, those proceeds can be many millions of dollars a year. "For these small towns in Wyoming, budgets aren't that big, so if you had a steady, reliable $5 million a year, that's a big deal."

People in communities benefitting from the financial windfall of new projects are grateful for those extra dollars this year. Throgmorton calls the wind industry "the goose that's laying the golden egg," and said he's relieved to be able to focus on his campus's development. And Lockman, the Cheyenne treasurer, said that even if the big tax increases from this year don't continue, it's been a boon during the pandemic months that will allow the city some leeway in 2021. "Thank goodness for wind energy," she said, "because if not for it, we'd be in a lot different shape."

EMILY ATKIN

Work from Home, Save the Planet? Ehhh

FROM Heated

BACK IN JUNE 2020—in the depths of pandemic hell—Heated wrote about the climate benefits of working from home.

We argued that when the pandemic was over, employers should continue the practice of allowing employees to work remotely if they desire. Our argument was based in part on a piece by Matt Butner and Jayni Hein, who said continuing flexible remote work policies could result in huge emission decreases while society transitions to renewable electricity-powered vehicles.

As the piece read: "No single activity contributes more greenhouse gas emissions than driving to and from work."

> Transportation is the number one source of emissions in the United States, and light-duty vehicles contribute the lion's share of all carbon emissions. The most common mile traveled by households operating light-duty vehicles is the one to and from work. . . .
>
> "Incentivizing these employees to work from home just two or three days a week will reduce their typical commute-based greenhouse gas emissions by half, immediately," Butner and Hein wrote. "This expeditious reduction in emissions would help immensely while the window to avoid catastrophic climate effects narrows."

One year later, the permanent shift to remote work appears to be underway. This could be profoundly beneficial for the U.S. greenhouse gas emission trajectory in the short-term.

But it could also be a huge debacle and a complete wasted opportunity if corporations and employees do not think of remote work policy as climate policy. And right now, they are definitely not thinking of it that way.

First: We're Not Going "Back to Normal" on Remote Work

In a survey of 231 large companies, *USA Today* reported that 79 percent plan to allow 10 percent or more of their employees to work remotely at least three days a week post-pandemic.

That's a huge increase from The Before, when only 29 percent of large companies allowed more than 10 percent of their employees to work from home semi-regularly.

In addition, 38 percent of the large companies surveyed said they will allow *40 percent or more* of their employees to work from home *all the time* post-pandemic.

That's a humungous difference from The Before, when only 5 percent of large companies allowed that.

Second: There's Huge Potential for Short-Term Climate Benefits

This shift in corporate remote work policy could have huge climate implications. Though it varies from city to city, the vast majority of Americans still drive to work. That means millions of cars are no longer going to be regularly on the road if these work-from-home policies continue. And I mean millions.

There are currently about 64.2 million office workers in the United States. If you apply the *USA Today* survey findings to the entire office workforce and assume workers are spread evenly across companies (they absolutely aren't, but whatever, just bear with me), then only a minimum of 1.8 million office workers were allowed to regularly telework before the pandemic. But now, approaching post-pandemic, a minimum of 4.9 million office workers will be allowed to regularly telework.

Three million people no longer regularly commuting to work is a huge climate deal. But *eight million people* no longer commuting to work *ever* is an even bigger climate deal.

That's the implication of the *USA Today* survey's second finding, according to my rudimentary analysis. Pre-pandemic, only 1.2 million people were allowed to work from home all of the time. And now, as we approach post-pandemic, it appears about 9.2 million people will be allowed to work from home all the time.

Reducing emissions from the transportation sector is one of the most difficult climate tasks we face. This increase in remote work could serve as a short-term way to address it as we transition to zero-emission vehicles.

Third: There's Also Huge Potential for This to Be Meaningless

This opportunity to address transportation emissions in the short-term through remote work is likely to fail, however, if companies and employees don't think of remote work policy as climate policy. If they simply think of remote work as a way to increase convenience and boost productivity, they will inevitably just replace driving emissions with other emissions and cancel the whole thing out.

Bloomberg explained this issue well back in March. If corporations aren't thinking of remote work as climate policy, they explained, then they might increase the number of trips for employees to meet one another face-to-face. That could "up the average number of airplane flights a year that the company is responsible for, handily erasing any emissions gains seen elsewhere."

In addition, if workers do not think of remote work as a climate-friendly act, then they may decide to take the opportunity to move away from cities. That will accelerate the "urban exodus" that's currently undermining the climate benefits of urban population density.

Remote work has also already led to a huge increase in household energy. That's bad, because an individual's household energy is far more likely to come from fossil fuel sources than a corporate office building's. (Corporations are increasingly under pressure to be more climate-friendly, and they're more likely to have the resources and time to make the shift.)

So, if both corporations and individuals aren't approaching remote work policy as a climate policy, there's huge potential for it to be a wasted opportunity. And spoiler alert: corporate America is currently thinking of it as pro-business policy. As *USA Today*'s article noted, corporations "increasingly believe remote workers have been cranking out more products and services." That's why this is happening. Not for the climate.

The Lesson: We're Not Going to Save the Planet by Accident

Politically, it's not considered smart to talk about climate policy as a thing that we need because it will save the planet and everything that lives on it. It's more politically safe to talk about climate policy as a thing that creates economic opportunity, and *it just so happens* that it saves the planet and everything that lives on it.

The prevailing logic for talking about climate policy like this is that somehow, it will trick Republicans into supporting the planet's long-term health. But actually, what it does is teach society that climate crisis is not a crisis worth addressing for its own sake.

The increase in corporate remote work policies post-pandemic could give the climate some much-needed short-term relief. But right now, it's just another way to fantasize that we can solve a crisis without ever having to be intentional about it.

SHIRA RUBIN

In Amsterdam, a Community of Floating Homes Shows the World How to Live Alongside Nature

FROM *The Washington Post*

MARJAN DE BLOK readjusts her body weight as she treads across the jetties linking a floating community on the river IJ. Her cheeks and nose are elfin red from the whipping winds. She shouts greetings to many of her neighbors, her voice carried by the water all around.

In October, heavy rains, hail, and fifty-mile-an-hour winds put Amsterdam on alert, just a short ferry ride away. But in the northern neighborhood of Schoonschip, life carried on mostly as usual. De Blok visited with neighbors to gossip and get updates on the local smart grid—which enables residents to generate and share energy with one another and the country—all while overhead lamps swayed and the homes glided up and down their steel foundational poles with the movement of the waters below.

"It feels like living at the beach, with the water, the saltiness of the air, and the seagulls," she says. "But it also feels special because, initially, we were told that building your own neighborhood, it's just impossible."

De Blok, forty-three, is a Dutch reality TV director by day and guerrilla sustainable commune organizer by night. She and her neighbors quickly adapted to life on water—proving, she says, that the technology already exists to make floating urban development a solution for the world's densely populated waterfront cities that are grappling with rising sea levels and the accelerating impacts of climate change.

Prince Harry, European lawmakers, and a long list of other dignitaries, urban planners, entrepreneurs, and citizens have come to Schoonschip in recent years, curious to see the real-life manifestation of a once sci-fi idea. On visitors' tours, De Blok has showcased Schoonschip's patchwork of environmentally focused social projects: lush floating gardens, tended by the residents and beloved by the waterbirds; a community center featuring floating architecture diagrams; and a nearby on-land vegetable patch bursting with kale in the winter and zucchini and tomatoes in the summer. But the homes' industrial-chic design and their immediate proximity to the city, De Blok says, are usually what surprise visitors most.

It's intentional, she says, as it helps to distinguish the dwellings from the quirky ten thousand converted barges—known as "houseboats"—that crowd the country's canals. Schoonschip, boasting modern design for modern lifestyles, seeks to serve as a prototype for the more than six hundred million people—10 percent of the world's population—who live on or near the water and are already being affected by climate change.

Accidental Pioneers

In the waterlogged Netherlands—a country that's a third below sea level and two-thirds flood-prone—floating homes are the latest in a centuries-long experiment in contending with water. Since the Middle Ages, Dutch farmer collectives have united to drain water to make room for agricultural land. The groups evolved into regional water boards that keep the land dry using a complex system of canals, dikes, dams, and sea gates. In 2007, the government unveiled a program called Room for the River, allowing certain locations to strategically flood during periods of heavy rain. Water management is such a normal part of Dutch discourse that many citizens are surprised to be asked about it, assuming it is common in every country. Dutch children as young as four are taught to swim with their clothes on, to instill "respect for the water," says Michiel Snijder, De Blok's partner, who works as a children's swimming instructor.

The Dutch have historically lived on water. As early as the seventeenth century, foreign tradespeople moored their boats to the land to sell their goods. In the 1960s, artists converted boats into homes to make houseboat living a culturally subversive way of opting out of civilization on land.

And especially as climate change has warmed the world's oceans over the past decade, Dutch water management strategists have sought to embrace, rather than resist, the rising sea levels. As part of that shift, floating communities have been emerging across Amsterdam, Rotterdam, and Utrecht. These homes that are converted into boats, rather than the other way around, bill themselves as part of a national, and potentially global, solution for a wetter future.

Schoonschip, home to about 150 residents that includes some 40 kids, is made up of 46 households located on 30 arks. Half are floating semidetached homes, shared by two families. One has three generations of the same family.

They are relatively low-tech, constructed off-site, and weighted by basins filled with recycled, water-resistant concrete, then pulled across the water by a tug and moored to the lake bed. Heavy pieces such as pianos are counterweighed with bricks on the opposite side of the house, and interior design is carried out in line with the Dutch principle of *gezellig,* or "coziness" (think: a Dutch version of hygge), which incorporates soft lighting, modern fixtures, and virtually no stylistic references to maritime life. Many rooms are outfitted with modular furniture that can be easily disassembled or reassembled to make room for life changes such as the birth of children or the separation of couples.

"Floating homes, you can turn them, flip them, take them with you. The flexibility on water is incomparable with the flexibility on land," says Sascha Glasl, a resident architect in Schoonschip. His architectural firm, Space & Matter, designed the community's jetty system and several of its homes. "It's evident that seawaters will rise, and that many big cities are really close to that water. It's amazing that not more of this innovation and building on water is being executed."

De Blok, who has no engineering, architecture, or hydrological training, says that she never intended to spearhead a movement in floating urban development.

In 2009, she was exhausted by living in Amsterdam. She was

working all the time, buying things she used just once or twice, and had very little time to meet with friends. She recycled and bought vintage instead of new but had the creeping feeling that she was being involuntarily made into a passive consumer.

On assignment on a cold winter day in 2009, she visited a solar-paneled floating event venue called geWoonboot as part of a series of short documentaries she was shooting on sustainable living. She was stunned by its contemporary feel, its immediacy to the water and the city, and its incorporation of experimental sustainability practices.

"Before I visited that boat, I wasn't really conscious that I didn't like the way I was living," she says.

When she asked friends if they had interest in building a floating community, she was unprepared for the deluge of responses. She cut off the list at 120 people, disappointing dozens.

She scouted waters around the geWoonboot neighborhood, known as Buiksloterham, a hundred-hectare, postindustrial area that had been largely abandoned since manufacturers—including the Shell oil company and the Fokker airplane factory that built parts for KLM airlines—left the city for lower-wage countries in the second part of the twentieth century.

"The area was a disaster, really depressing. Just some companies, no streetlights," De Blok recalls.

But when she got a look at the city's plans to develop tens of thousands of housing units and cultural centers in the area, she thought, "We could be pioneers here."

"*Schoonschip*" means "clean ship," which when made into a verb, "to do schoonschip," means "starting over from scratch." In Buiksloterham, the twenty-two-story Shell tower has been rebranded as the Amsterdam Dance and Music Tower, with dance clubs, a revolving restaurant, and an observation deck. The grassy Overhoeks Promenade, which served as a gallows from the fifteenth to the eighteenth century, hosts the hulking, modernistic Eye Film Museum. The NDSM wharf is peppered with artist collectives, vintage shops, and a luxury hotel atop the world's tallest harbor crane.

De Blok views water as much in engineering as in social terms, especially as densely populated cities such as Amsterdam undergo rapid gentrification, replacing social housing and middle-class neighborhoods with homes for the ultrarich and Airbnbs for tourists.

Looking to make Schoonschip something different, she had all residents sign a manifesto committing them to constructing, insulating, and finishing their homes with eco-friendly materials such as straw, burlap, and bamboo. They also informally signed up for eating together, swimming in their "backyards" together, and conducting their lives largely in common view of one another, with curtains only rarely drawn. They share bikes and cars and use a vibrant WhatsApp group to request almost any service or borrow virtually any item from neighbors, which they can have delivered to their doorstep usually within a few minutes. Every Tuesday, many of the residents order two-course vegan meals prepared by a resident chef, which they often share in one another's homes.

The neighborhood feels like an extended block party mostly because many of the residents are actually De Blok's friends, or friends of friends, including many colleagues from the TV and entertainment industry. There's a celebrity talk show host, several heads of content, and a podcaster, most of whom joined the project in their twenties and thirties, when they had no kids and ample time to invest in building a community from scratch. Twelve years of bureaucratic struggles later, those young single couples are young families. During the summer months, their children jump out of their bedroom windows directly into the water below. On clear winter nights, the neighborhood gleams with soft lighting and buzzes with the hum of chattering residents, parked out on their top-floor porches where they have a front-row view to the inky water and the starry sky.

"When it's dark and all the lights in the houses are on, it feels like a set from a film," De Blok says.

To realize Schoonschip's sustainability goals, De Blok needed to draw from its most valuable and multipurpose resource: the residents themselves. Siti Boelen, a Dutch television producer, mediated between the Schoonschip representative committee and the local municipality. Glasl, the architect, helped design the five rows of jetties that connect each house to the others and to the land.

Eelke Kingma, a resident and renewable tech expert, joined a community task force that received special permission from the experimental sector of a Dutch electricity company to design the neighborhood's smart grid system. Residents collect energy from five hundred solar panels—placed on roughly a third of the community's roofs—and from thirty efficient heat pumps that draw from the water below. They then store this energy in enormous batteries located beneath their homes and sell any surplus to one another, as well as to the national grid.

Kingma, with parnters, is finishing a new AI-automated program that will use the homes' smart meters to inform residents when they can earn the most from selling, based on the fluctuations in energy market prices. This would make Schoonschip the first neighborhood in the country to turn a profit from generating energy, Kingma says. It's made possible by the fact that each home in Schoonschip has five to eight smart meters—most homes in the Netherlands have only one—which constantly track the influx and outflow of the underwater energy storage system.

The program is being monitored in collaboration with fifteen European companies, universities, and institutions, organized by the European Commission, which supports renewable energy experiments in the hopes of scaling them up across the continent.

Requests from All Over the World

Over the past decade, the floating-house movement has been gaining momentum in the Netherlands.

The Dutch government is amending home-owning laws to redefine floating homes as "immovable homes" rather than "boats," to simplify the process of obtaining permits.

"Building on water is considered a kind of blank canvas: due to the lack of existing infrastructure," reads a research paper that advocates for the amendment to the law. "We foresee that in the near future building on water and floating living in the Netherlands will no longer be a luxury, but an absolute necessity."

Amsterdam and Rotterdam, the Dutch delta city located 90 percent below sea level, are reporting a sharp uptick in requests for permits to build on the water. The trend is coinciding with a national water awareness campaign for an era in which climate change is already a fact of life. The government launched an app called Overstroom Ik?, or Will I flood?, allowing residents to check if their zip code is at risk of flooding, and supported a traveling pop-up art installation called Waterlicht, or Waterlight, that for six years has been projecting blue lights over New York, Dubai, and many other of the world's largest cities to simulate a virtual flood.

In Room for the River program areas, low-lying parks and beaches will feature public awareness campaigns during the non-flood season, highlighting water-related problems and potential solutions.

And as they expand, the initiators of Schoonschip and other floating neighborhoods, office buildings, and event spaces across the Netherlands are increasingly being consulted for projects across the world.

The potential solution has grown in prominence as sea levels are forecast to rise three to five feet this century, and storms are expected to increase in frequency and intensity. Last summer, at least 220 people died in Germany and Belgium from a once-in-four-hundred-year rain event. In China, nearly eight inches of rain fell in one hour. New York City recorded its fifth wettest day on record. Rivers submerged parts of Tennessee that were not previously considered floodplains.

By the end of this century, the kind of intense precipitation events that would typically occur two times per century will occur twice as often, and events that would occur once every two hundred years will become four times as frequent, according to research published in August by a team of water experts led by hydrologist Manuela Brunner.

Marthijn Pool, the cofounder of Space & Matter, has been among the growing number of Dutch architects exporting their knowledge of floating architecture to the United States. Pool says that land-scarce New York is especially ripe for floating development, and that awareness of the need for innovative solutions has grown in the aftermath of Hurricane Sandy, which inundated New York City with waves as high as fourteen feet in 2012. Space & Matter is planning to build a research and development center in an area off Red Hook, with hulls designed to promote kelp oysters underneath. They are also showcasing designs for a second residential project, similar to Schoonschip, in Washington, D.C., which Pool hopes can set an example for stormproofing waterfront communities, as well as create an economic incentive for floating architecture.

"Water is relatively cheap, you don't need to put sewage in the sidewalks and if you can make your own grid, then you're able to provide your own infrastructure," says Pool, adding that the homes can save cities billions of dollars in damages since they can withstand high levels of precipitation, simply moving up when waters rush in and descending to their original position when waters recede. He says that sewage, if connected to an independent sustainable sanitation system like that being developed by Schoonschip, can be separated and used for irrigating plants.

Koen Olthuis, an architect with the floating architectural firm Waterstudio, which designed several of the houses in Schoonschip and in the nearby floating Amsterdam neighborhood of Ijburg, says that an increase in organizations and governments looking to adapt to climate change has introduced floating projects to places that may need it most. In 2013, his firm sent a floating, internet-connected converted cargo container, called *City App*, to the Korail Bosti slum of Dhaka, Bangladesh, where children used the space during the day to attend remote classes, and adults used it to develop business projects at night. In 2019, the vessel was relocated to a slum near Alexandria, Egypt, where it remains stationed. He says that the project, along with his firm's luxury floating villas built in countries like the United Arab Emirates, are paving the way to mainstream the concept of mobile floating homes, by which homes could unmoor from lakebeds for calmer waters in the case of extreme flooding or even unpleasant weather.

"We want to upgrade cities near the water," he says. "Now we're at a tipping point where it's actually happening. We're getting requests from all over the world."

After two decades of planning, his firm will oversee construction on a new two-hundred-hectare lagoon off Malé, the capital of the Maldives. The city sits less than three feet above sea level, making it vulnerable to even the slightest rise. The small, simply designed housing complex, priced at around 10 to 15 percent more than comparable houses on land, is intended for twenty thousand people. It will have water pumps that draw energy from deep-sea water, a water-based city grid, and homes with artificial coral-clad hulls to encourage marine life.

"Today, we can see how cities are performing with floods, extreme weather, urbanization," says Olthuis, adding that the initial Dutch projects and now-international iterations are showing that "we can cope with the challenges of sea level rises."

In Schoonschip, De Blok says that she hopes everyone will have the opportunity to live on water one day.

"It does something to you, being aware that under your house everything is moving," she says. "There's some magic to it."

JESSICA PLUMB

A River Reawakened

FROM *Orion*

IN SEPTEMBER 2011, I stood on a river overlook with children from my daughter's elementary school, all of us transfixed by a giant jackhammer pounding cement to rubble. Below us, a waterfall raged through the first notch carved in the Lower Elwha Dam, as dust rose in the September sunshine, drifting over Douglas fir and cedar crowns. Trees were the only spectators old enough to remember when the Elwha River ran free, a century earlier. The rest of us stood in awe, watching the world's largest dam removal to date, feeling time start to spin in reverse.

I've spent a decade bearing witness to an unprecedented restoration experiment in Washington State. That September day committed me to unraveling the river's story, while dam removal raised enough questions to keep scientists engaged for years to come.

The Elwha River begins in the Olympic Mountains, a rugged range encircled by ocean on three sides. Rivers radiate from the center of the Olympic Peninsula, short and steep, born from deep snowpack at the heart of the range. The Elwha River tumbles over four thousand feet in forty-five miles, from alpine meadows into rock canyons and floodplains, until it meets the sea in the Strait of Juan de Fuca.

When the first dam rose in 1911, it blocked the river just five miles upstream from the ocean. A second dam, constructed in Glines Canyon shortly before the creation of Olympic National Park, would operate within park boundaries for more than seven decades.

The beauty and diversity of Olympic National Park draws visitors from around the world. Many come to experience its "ancient groves" of old-growth forest. To me, this familiar landscape does not feel ancient. I imagine this is what the world was like when it was young. Freshly scrubbed by glaciers, with terrain like a restless teenager, prone to earthquakes and pulsing with life. The first time I took my daughter to a salmon creek during a fall run, we heard fish long before we could see them, splashing and slapping their tails in their raucous run upstream. Next came the stench of carcasses, salmon that had spawned and died, lining the bank. Finally, we caught a glimpse of the creek, where a cloud of pink salmon stirred an azure pool, circling below a stretch of rapids. Pink salmon arrive every other year, odd years only. When they return, they are the wildebeest of the Olympic Peninsula, a migration that once felt uncountable.

Wild salmon weave freshwater and saltwater ecosystems together, and rivers are the threads between those worlds. Impounding a river severs that connection in both directions. When dams were built without fish passage on the Elwha River, the most immediate impact was a barrier for anadromous fish. The river had legendary fish runs, attracting all five species of Pacific salmon, unusual in the region. The most iconic species was Chinook, famous for its enormous size. Members of the Lower Elwha Klallam Tribe describe hundred-pound salmon, fish so big that children struggled to drag a single one home in a gunnysack.

A century after the first dam was constructed, these giant Chinook lived on in memory, with a remnant population supported by a hatchery. Pink salmon teetered on the edge of extirpation. Numbers of all anadromous fish in the Elwha, ten species including salmon, steelhead, and trout, had dwindled to a tiny fraction of the river's past abundance. To visitors, the Elwha Valley felt wild, with much of the watershed protected by Olympic National Park. People who knew the river well heard the silence and saw a system on the verge of collapse.

Dam removal began when my daughter was in kindergarten, after decades of advocacy by the Lower Elwha Klallam Tribe, environmental groups, and politicians. Now I trek into the valley with a teenager, hiking around washouts on a road we once drove. As we head upriver, a bobcat slinks across remaining pavement and disappears into a stand of big-leaf maple. Low mist hangs over the valley, as we orient ourselves in a changing landscape. When the upper dam came down, the river began to wander and tear at an access road that served a power station and the national park, carving unfamiliar channels with every flood.

We return each autumn to watch for salmon, the headline success story of dam removal, but this trip is timed for a different annual ritual. When the takedown of Elwha dams shifted from a crazy idea to imminent reality, revegetation specialists planned for the moment when water would drain away from two reservoirs, exposing land buried for a hundred years. An estimated 24 million cubic yards of sediment collected behind the dams, creating a challenge for ecologists who hoped to jump-start the process of plant succession. Their goal was to outrace invasive plant species in eight hundred acres laid bare by dewatering, and to set the stage for a future forest. Crews collected seeds from the Elwha Valley long before the dams came down, propagating native plants by the thousands. Volunteers transplanted more than three hundred thousand starts while teams prepared to scatter tons of natively sourced seed. As the final day approached, restoration ecologist Joshua Chenoweth added one last seed species to the master plan, despite limited supply: lupine, known for its nitrogen-fixing capacity.

Ten years later, the trail into the former reservoir, once known as Lake Mills, begins at an abandoned boat ramp. Our family pushes through a young forest of cottonwood and willow saplings, crowding the ramp where kayaks once launched. We call out as we enter a dense thicket of growth, hoping to alert black bears. The sweet smell of lupine reaches us before we emerge from the willows, into a sea of purple. The former reservoir is awash in lupine. These flowers are the first step in a slow transformation from stranded sediment to mature forest.

When salmon disappear, an entire watershed begins to change in subtle ways. Fish went missing from the whisper of trees, changing the inner workings of the forest. When salmon return to mountain streams, they transport marine-derived nutrients deep inland, feeding the forests that shade their spawning habitat. Scientists have shown the presence of ocean nutrients in the tissue of trees, many miles from shore. Migrating salmon attract wildlife, from otter to bear, who in turn help to distribute carcasses throughout the forest.

The single biggest question surrounding Elwha Dam removal was: Would the fish return, after a hundred-year absence? Fish numbers have increased steadily since the dams came down, and by 2019 Elwha Chinook numbered well over seven thousand adults. Mike McHenry, fisheries biologist at the Lower Elwha Klallam Tribe, says that 2019 numbers suggest a fivefold increase for juvenile Chinook, the first "strong signal" for natural Chinook production in newly available habitat. He notes that river conditions began to stabilize in recent years. During initial salmon seasons after dam removal, turbid water and shifting channels posed a challenge for returning fish. During floods, the river churned the color of chocolate milk, creating a sediment plume visible offshore, transforming the estuary and nearshore habitat. As these sediment pulses began to subside, the river and estuary came to life.

In addition to fish numbers, biologists are tracking the distribution and life histories of fish. A surge of summer steelhead in the upper river surprised scientists, who began to suspect that genetically similar rainbow trout were rediscovering anadromy—a capacity they had stilled for generations. Studies on Elwha steelhead continue, but initial findings suggest a remarkable story, one in which genetic potential outlasted many generations of entrapment. Anadromous species have ventured well beyond both former dam sites, with steelhead and Chinook traveling the farthest upstream. Throughout the watershed, scientists are observing diversity and changes in life strategy spurred by dam removal. Bull trout numbers have doubled, as they venture downstream into the estuary. Coho salmon rediscovered a lake previously walled off by dams, where their smolt are thriving.

The rewilding of the Elwha is a story of environmental justice, equal in scope to the scale of restoration. Salmon are at the heart of the Lower Elwha Klallam Tribe's culture and economy, and tribal members were the first to protest the degradation of the river. As water rose behind the first dam, the reservoir inundated a site sacred to the tribe, believed to be the birthplace of the Klallam people. When the reservoir drained away, tribal members rediscovered the sacred place they'd lost a century earlier. This memory is now a physical place, visited by tribal members. The Lower Elwha Klallam Tribe continues to play a leading role in the restoration effort, working with Olympic National Park and numerous agencies to track and foster habitat recovery.

Tribal land encompasses the east side of the estuary, and for the first time in a century, the tribe is gaining, rather than losing, land. Rivers move mountains to the sea. On a dammed river, sediment collects where water slows, filling reservoirs with gravel as the shoreline shrinks. Tribal elders remember harvesting shellfish near the estuary; however, cobblestones replaced this productive habitat over time, due to sediment trapped upstream. When the dams fell, these deposits began to flood downstream, rebuilding sandbars at the river's mouth. The changing habitat quickly drew Dungeness crab and other species back to the near shore. Geologists tracking the growing shoreline have noted an unexpected benefit of dam removal: new sand deposits will help to buffer the shoreline against the impact of sea level rise.

Perhaps the most astonishing element of the Elwha restoration is how quickly ecosystem changes can be reversed. From sandbars stretching into the Strait of Juan de Fuca to changes in wildlife upstream, the whole watershed has responded more rapidly than expected. With salmon numbers growing steadily, scientists are tracking species associated with migrating fish. One study considered American dippers, the West's only aquatic songbird. As salmon returned to the Elwha, these charismatic birds were found to have marine-derived nutrients in their diet, which in turn made them more likely to stay in salmon-rich territory year-round, and to double-brood in one season. The Elwha experiment is revealing the role of salmon in countless other species, terrestrial and aquatic. Taken together, studies throughout the valley underscore a simple message: Everything is connected. *And with connection, comes life.*

On the eve of dam removal in 2011, Senator Bill Bradley traveled across the country to speak about "the great gift of the Elwha: hope." September 2022 marks a decade since speeches gave way to the hard work of dam removal and river restoration. Hope in 2021 can feel elusive, particularly amid a season of climate disruption. Yet the first ten years of the Elwha experiment have been breathtaking in speed and scope, a testament to the power of nature and potential for renewal.

Olympic National Park left the buttresses of Glines Canyon Dam intact, as a monument to what was, and what can be. Today the remains of the dam offer a dizzying view. Two hundred feet down, amid jagged rocks, a river of light threads the canyon. Wind moves through the notch of the former dam, channeled by the valley. With wind on my face, and white water coursing below, I feel something beyond hope: fleeting pure joy, mixed with gratitude for the chance to bear witness. I long to speak to the river that has roared and whispered in my head for so many years, to ask the river a question: Is this what the world was like when it was young?

SOMINI SENGUPTA, CATRIN EINHORN,
AND MANUELA ANDREONI

There's a Global Plan to Conserve Nature. Indigenous People Could Lead the Way

FROM *The New York Times*

WITH A MILLION SPECIES at risk of extinction, dozens of countries are pushing to protect at least 30 percent of the planet's land and water by 2030. Their goal is to hammer out a global agreement at negotiations to be held in China later this year, designed to keep intact natural areas like old-growth forests and wetlands that nurture biodiversity, store carbon, and filter water.

But many people who have been protecting nature successfully for generations won't be deciding on the deal: Indigenous communities and others who have kept room for animals, plants, and their habitats, not by fencing off nature, but by making a small living from it. The key to their success, research shows, is not extracting too much.

In the Brazilian Amazon, Indigenous people put their bodies on the line to protect Native lands threatened by loggers and ranchers. In Canada, a First Nations group created a huge park to block mining. In Papua New Guinea, fishing communities have set up no-fishing zones. And in Guatemala, people living in a sprawling nature reserve are harvesting high-value timber in small amounts. In fact, some of those logs could end up as new bike lanes on the Brooklyn Bridge.

"If you're going to save only the insects and the animals and not the Indigenous people, there's a big contradiction," said José Gregorio Díaz Mirabal, who leads an umbrella group, the Coordinator of Indigenous Organizations of the Amazon River Basin. "We're one ecosystem."

Nature is healthier on the more than quarter of the world's lands that Indigenous people manage or own, according to several scientific studies. Indigenous-managed lands in Brazil, Canada, and Australia have as much or more biodiversity than lands set aside for conservation by federal and other governments, researchers have found.

That is in stark contrast to the history of conservation, which has a troubled record of forcing people off their land. So, it is with a mixture of hope and worry that many Indigenous leaders view this latest global goal, known as 30×30, led by Britain, Costa Rica, and France. Some want a higher target—more than 50 percent, according to Mr. Díaz Mirabal's organization—while others fear that they may once again be pushed out in the name of conservation.

Defending Land, Protecting Vital Forests

In the Brazilian Amazon, Awapu Uru-Eu-Wau-Wau puts his life on the line to protect the riches of his ancestral lands: jaguars, endangered brown woolly monkeys, and natural springs from which seventeen important rivers flow. His people, the Indigenous Uru-Eu-Wau-Wau, have legal right to the land, but must constantly defend it from armed intruders.

Just beyond their seven-thousand-square-mile territory, cattle ranchers and soy planters have razed much of the forest. Their land is among the last protected forests and savanna left in the Brazilian state of Rondônia. Illegal loggers often encroach.

So, Mr. Uru-Eu-Wau-Wau, who uses his community's name as his surname, patrols the forest with poison-tipped arrows. Others in his community keep watch with drones, GPS equipment, and video cameras. He prepares his daughter and son, eleven and thirteen years old, to defend it in the years ahead.

"No one knows what's going to happen to us, and I'm not going to live forever," Mr. Uru-Eu-Wau-Wau said. "We need to leave it to our children to get on with things."

The risks are high. Mr. Uru-Eu-Wau-Wau's cousin, Ari Uru-Eu-Wau-Wau, was murdered last April, part of a chilling pattern among land defenders across the Amazon. In 2019, the most recent year for which data is available, at least forty-six were murdered across Latin America. Many were Indigenous.

The community's efforts have outsized benefits for the world's 7.75 billion people: the Amazon, which accounts for half the remaining tropical rainforest in the world, helps to regulate Earth's climate and nurtures invaluable genetic diversity. Research shows Indigenous property rights are crucial to reducing illegal deforestation in the Amazon.

A Collapse of Nature

Nature is under assault because humans gobble up land to grow food, harvest timber, and dig for minerals, while also overfishing the oceans. Making matters worse, the combustion of fossil fuels is warming up the planet and making it harder for animals and plants to survive.

At fault, some scholars say, are the same historical forces that have extracted natural resources for hundreds of years, at the expense of Indigenous people. "What we're seeing now with the biodiversity collapse and with climate change is the final stage of the effects of colonialism," said Paige West, an anthropologist at Columbia University.

There is now broad recognition that reversing the loss of biodiversity is urgent not only for food security and a stable climate, it's also critical to reducing the risk of new diseases spilling over from wild animals, like the coronavirus.

Enter 30×30. The goal to protect at least 30 percent of the Earth's land and water, long pushed by conservationists, has been taken up by a coalition of countries. It will be part of diplomatic negotiations to be held in Kunming, China, this fall, under the United Nations Convention on Biodiversity. The United States is the only country, apart from the Vatican, that has not joined the convention, though President Biden has ordered up a plan to protect 30 percent of American waters and lands.

Indigenous communities are not recognized as parties to the international agreement. They can come as observers to the talks but can't vote on the outcome. Practically though, success is impossible without their support.

They already protect much of the world's land and water, as David Cooper, deputy executive secretary of the United Nations agency for biodiversity, pointed out. "People live in these places," he said. "They need to be engaged and their rights respected."

A coalition of Indigenous groups and local communities has called for the agreement to protect at least half of the planet. Scientific research backs them up, finding that saving a third of the planet is simply not enough to preserve biodiversity and to store enough planet-warming carbon dioxide to slow down global warming.

Creating a New Kind of Park

A half century ago, where boreal forest meets tundra in Canada's Northwest Territories, the Łutsël K'é' Dene, one of the area's Indigenous groups, opposed Canada's efforts to set up a national park in and around its homeland.

"At that time, Canada's national parks policies were very negative to Indigenous people's ways of life," said Steven Nitah, a former tribal chief. "They used to create national parks—fortress parks, I call it—and they kicked people out."

But in the 1990s, the Łutsël K'é' Dene faced a new threat: diamonds were found nearby. They feared their lands would be gutted by mining companies. So, they went back to the Canadian government to revisit the idea of a national park—one that enshrined their rights to manage the land, hunt, and fish.

"To protect that heart of our homeland from industrial activities, this is what we used," said Mr. Nitah, who served as his people's chief negotiator with the Canadian government.

The park opened in 2019. Its name, Thaidene Nëné, means "Land of the Ancestors."

Collaboration among conservationists, Indigenous nations, and governments holds a key to protecting biodiversity, according to research.

Without local support, creating protected areas can be useless. They often fail to conserve animals and plants, becoming so-called "paper parks."

Making a Living from Nature

Researchers have found that biodiversity protection often works best when local communities have a stake.

On islands in Papua New Guinea, for example, where fish is a staple, stocks had dwindled in recent decades. Fishers ventured farther from shore and spent more time at sea but came back with smaller catches. So, they partnered with local and international nonprofit groups to try something new. They changed their nets to let smaller fish escape. They reduced their use of a poison that brings fish to the surface. Most critically, they closed some waters to fishing altogether.

Meksen Darius, the head of one of the clans using these measures, said people were open to the idea because they hoped it would improve their livelihoods.

It did.

"The volume, the kinds of species of fish and other marine life, they've multiplied," Mr. Darius, a retired lawyer, said.

Recent research from around the world shows that marine protected areas increase fish stocks, ultimately allowing fishing communities to catch more fish on the edges of the reserves.

To Iliana Monterroso, an environmental scientist at the Center for International Forestry Research in Lima, Peru, what matters is that people who live in areas of high biodiversity have a right to manage those areas. She pointed to the example of the Mayan Biosphere Reserve, a territory of two million hectares in Guatemala, where local communities have managed the forest for thirty years.

Under temporary contracts with the national government, they began harvesting limited quantities of timber and allspice, selling ornamental palms, and running tourism agencies. They had an investment to protect. "The forest became the source of livelihood," Dr. Monterroso said. "They were able to gain tangible benefits."

Jaguars, spider monkeys, and 535 species of butterflies thrive there. So does the white-lipped peccary, a shy pig that tends to disappear quickly when there's hunting pressure. Community-managed forests have fewer forest fires, and there is almost zero rate of deforestation, according to researchers.

Erwin Maas is among the hundreds of Guatemalans who live

there, too. He and his neighbors run a community-owned business in the village of Uaxactún. Mahogany is plentiful, but they can take only so much. Often, it's one or two trees per hectare per year, Mr. Maas said. Seed-producing trees are left alone.

"Our goal is to sustain ourselves with a small amount and always take care of the forest," he said.

Contributors' Notes

*Other Notable Science and
Nature Writing of 2021*

Contributors' Notes

JULIAN AGUON is an Indigenous human rights lawyer and writer from Guam. He is the author of *No Country for Eight-Spot Butterflies,* based on his widely acclaimed debut, *The Properties of Perpetual Light.* His essay "To Hell with Drowning" was a finalist for the 2022 Pulitzer Prize for Commentary.

MANUELA ANDREONI is a writer for the Climate Forward newsletter. Before joining the climate desk, she was a fellow at the Rainforest Investigation Network, covering the Brazilian Amazon. She studied at the Federal University of Rio de Janeiro and received a master's degree from Columbia University, where she was a fellow at the Toni Stabile Center for Investigative Journalism. She is a native of Rio and started writing for the *Times* in 2018 working out of the Brazil bureau.

MARK ARAX is a two-time winner of the California Book Award and the William Saroyan International Writing Prize from Stanford University. He is the author of several books, including his most recent national bestseller *The Dreamt Land: Chasing Water and Dust Across California.*

EMILY ATKIN is the founder of Heated, a newsletter dedicated to accountability reporting and analysis on the climate crisis. She is also a contributing essayist to the climate solutions anthology *All We Can Save* and an enthusiastic but deeply amateur drummer.

JUSTINE CALMA has reported on climate change across four continents since the adoption of the Paris Agreement in 2015. Now she covers energy and the environment as a science reporter at Vox Media's the Verge. Calma moved from the Philippines to California as a kid and is now based in New York City.

CATRIN EINHORN is a journalist who covers biodiversity, climate, and the environment for the *New York Times*. She was part of a team of reporters that received the Pulitzer Prize for Public Service in 2018 for exposing sexual harassment and misconduct in the American workplace. Throughout her career, Einhorn has focused on narrative-driven work in print, film, and audio. Her documentary *Father Soldier Son*, directed and produced with Leslye Davis, won an Emmy Award.

YESSENIA FUNES is an environmental journalist covering the climate crisis through an intersectional lens. She's the climate director for *Atmos*, a nonprofit magazine dedicated to covering climate and culture. She's been published in the *Guardian*, HuffPost, *Vogue*, *Cosmopolitan*, Gizmodo, Grist, and more.

CLAUDIA GEIB is a science journalist based on Cape Cod in Massachusetts, covering marine science, wildlife, and the environment. She is the producer of the podcast *Gastropod* as well as the author of a forthcoming book on elephants. When not writing, you can often find her near, on, or under the water.

JEFF GOODELL's most recent book, *The Water Will Come: Rising Seas, Sinking Cities, and the Remaking of the Civilized World*, was a *New York Times* Critics Top Book of 2017. He is the author of five previous books, including *Big Coal: The Dirty Secret Behind America's Energy Future*, and a contributing editor at *Rolling Stone*, where he has covered climate change for more than a decade. He is a Senior Fellow at the Atlantic Council and a 2020 Guggenheim Fellow.

CYNTHIA R. GREENLEE, PH.D., is a North Carolina–based historian, writer, and editor. Her work has appeared in publications as diverse as *The Atlantic*, *The Best American Food Writing*, *Essence*, Longreads, *The Nation*, the *New York Times*, *Smithsonian*, and Vox. She is also a winner of a James Beard Award for excellence in food writing. Check out her work at cynthia greenlee.com.

MIKKI K. HARRIS is a multimedia journalist whose work over the past twenty years has focused on community-based power as examined through the tools of journalism and cultural studies. She uses oral history, documentary photojournalism, and video, as well as writing, to document and shape stories of impact. Harris became a certified drone pilot and founded the Atlanta Drone Lab in 2019. In addition to drone technology, she produces stories in virtual reality while designing and teaching digital media and visual innovation courses at Morehouse College.

ʻCÚAGILÁKV (JESS ḢÁUSṪI) is a mother, writer, and community organizer from the Heiltsuk Nation in Bella Bella, British Columbia. She is a poet, plant worker, and nonprofit leader helping Heiltsuk youth and families learn and thrive on the land. She lives in her unceded ancestral homelands with her extended family and nonhuman kin.

KASHMIR HILL is a tech reporter at the *New York Times.* She writes about the unexpected and sometimes ominous ways technology is changing our lives, particularly when it comes to our privacy. She is currently working on a book about facial-recognition technology.

BOB HOLMES trained as an evolutionary biologist and has had a long career as a science writer, first for *New Scientist* magazine and more recently for Knowable Magazine. He lives in Edmonton, Alberta.

JANE C. HU is an independent journalist living in Seattle.

SABRINA IMBLER is a writer living in Brooklyn, New York. Their writing has appeared in the *New York Times, The Atlantic,* and *Sierra.* Their first book, *How Far the Light Reaches,* will be published in December 2022.

COREY G. JOHNSON was an investigative reporter at the *Tampa Bay Times* from January 2017 to May 2022. His work on "Poisoned" with Rebecca Woolington and Eli Murray won the Pulitzer Prize for Investigative Reporting, the George Polk Award, the IRE Gold Medal, and several other national honors. Years prior, his work for the Center for Investigative Reporting exposed hundreds of illegal and coerced sterilization surgeries of women prisoners. He prompted sweeping reforms in California, including a statewide ban and the first-ever reparations program for victims and families. His investigation of public schools uncovered shocking statewide failures in the construction of earthquake protections. He was born and raised in Atlanta and is a proud graduate of Florida A&M University in Tallahassee. He works for ProPublica, a nonprofit news organization based in New York City.

LACY M. JOHNSON is a Houston-based professor, curator, and activist, and the author of the essay collection *The Reckonings,* the widely acclaimed memoir *The Other Side,* and *Trespasses.* She is also the editor, along with graphic designer Cheryl Beckett, of *More City Than Water: A Houston Flood Atlas.* She teaches creative nonfiction at Rice University and is a founding director of the Houston Flood Museum.

SARAH KAPLAN is a *Washington Post* reporter covering the unequal impacts of climate change and humanity's response to a warming world.

ARIANNA S. LONG, PH.D., is a NASA Hubble postdoctoral fellow at the University of Texas, Austin, and former Ford fellow at the University of California, Irvine. Long uses state-of-the-art telescopes to understand how massive galaxies form, grow, and die in the early cosmos.

CHRIS MALLOY is a writer from the American Northeast now based in the Southwest. For stories, he has gone foraging in the Rockies and Sonoran Desert, trailed a quest to brew ale overnight in a forest, roamed a lost mountain orchard with cider-making scientists, and gone hunting for *gloscho* with Apaches. He covers mostly food, culture, agriculture, the environment, and innovation.

ELI MURRAY is an investigative reporter who specializes in data work at the *Tampa Bay Times*. He's a self-taught programmer who uses code to create graphics and crunch numbers on statewide and national investigations. He recently worked as part of a reporting team with Corey G. Johnson and Rebecca Woolington that exposed dangerous working conditions at a Florida lead factory. The series was the recipient of the 2022 Pulitzer Prize for Investigative Reporting, a George Polk Award, and several other national awards. Murray grew up in rural Illinois and received an associate's degree from Sauk Valley Community College before pursuing journalism at the University of Illinois. He joined the *Tampa Bay Times* in 2015.

KENDRA PIERRE-LOUIS is a senior climate reporter with the Gimlet/Spotify climate solutions podcast *How to Save a Planet*. Previously she was a climate reporter with the *New York Times* and a staff writer for Popular Science, where she wrote about science, the environment, and, occasionally, mayonnaise. In addition, her writing has appeared in *The Atlantic,* the *Washington Post,* and Slate. She is based in Queens, New York.

JESSICA PLUMB is an award-winning filmmaker and writer, known for exploring the relationship between people and place. She is the producer, codirector, and writer of the feature documentary *Return of the River,* chronicling the world's largest dam removal to date on the Elwha River. Her essays can be seen in *Orion,* the *Seattle Times,* and *Mountaineer,* among others, and her writing for film has been recognized as the "best writing in science media" by the Jackson Hole Science Media Awards. Plumb is based on the Olympic Peninsula, near the wilderness and waters she loves.

RACHEL RAMIREZ is a writer at CNN, covering all things climate change and environmental justice. She is currently based in New York City but was born and raised on the Pacific island of Saipan in the Northern Mariana Islands, a U.S. territory just north of Guam. You can find her previous work in Vox, HuffPost, Grist, and other publications.

RUTH ROBERTSON (formerly Hopkins) is a Dakota/Lakota Sioux writer and enrolled member of the Sisseton Wahpeton Sioux Tribe. She is also a biologist and the chief judge of her tribe. She was born on the Standing Rock Indian Reservation and resides on ancestral treaty lands.

JULIA ROSEN is an independent journalist covering science and the environment. She writes about how the world works and how humans are changing it. Her work has appeared in the *New York Times, The Atlantic, Scientific American,* and *High Country News,* among other publications, and she is a former science reporter for the *Los Angeles Times.* She lives in Portland, Oregon, with her family.

SHIRA RUBIN is a correspondent for the *Washington Post* based in Tel Aviv, Israel, where she covers conflict, climate, technology, and health care in the Middle East and Europe. She's on Twitter at @shira_rubin.

SOMINI SENGUPTA is an international climate correspondent, the lead writer of the *New York Times*' Climate Forward newsletter, and the author of *The End of Karma: Hope and Fury Among India's Young* and has covered the Middle East, West Africa, and South Asia for the *Times.* She received a George Polk Award for international reporting and awards from the Overseas Press Club, the Newswomen's Club of New York, and the National Association of Black Journalists.

SONIA SHAH is a science journalist and prize-winning author of critically acclaimed books on science, politics, and human rights. Her latest book, *The Next Great Migration: The Beauty and Terror of Life on the Move,* explores our centuries-long assumptions about migration through science, history, and reporting, predicting its lifesaving power in the face of climate change. A finalist for the 2021 PEN/E. O. Wilson Literary Science Writing Award, it was selected as a best nonfiction book of 2020 by *Publishers Weekly,* a best science book of 2020 by Amazon, a best science and technology book of 2020 by *Library Journal,* and a Tata Literature Live! finalist for the best book of the year.

LUCY SHERRIFF is a British multimedia journalist based in Los Angeles. She focuses on climate and social justice, culture, and the environment, and is working on a podcast about Indigenous communities and their relationship with the land. Her most recent work is directing a documentary on women who bring up their children in prison.

LISA SONG is a reporter at ProPublica who covers climate change and the environment. She previously worked at Inside Climate News, where she helped reveal Exxon's shift from conducting early global warming research to supporting climate denial.

MEERA SUBRAMANIAN is an award-winning independent journalist, author of *A River Runs Again: India's Natural World in Crisis,* and a contributing editor of *Orion* magazine. Based on a glacial moraine on the edge of the Atlantic, she's a perpetual wanderer who can't stop planting perennials. You can find her at meerasub.org.

JAMES TEMPLE is the senior climate and energy editor at *MIT Technology Review,* where he covers technologies and policies designed to cut emissions, draw down carbon, or otherwise address rising climate risks. He was previously a senior director at the Verge, deputy managing editor at Recode, and columnist at the *San Francisco Chronicle.*

LISA WELLS is the author, most recently, of *Believers: Making a Life at the End of the World,* a finalist for the 2022 PEN/E.O. Wilson Literary Science Writing Award. Her writing has appeared in *Harper's, Granta, n+1,* and the *New York Times,* and she writes a regular column for *Orion* called Abundant Noise. She lives in Seattle.

REBECCA WOOLINGTON is the investigative editor at the *Tampa Bay Times.* She was previously an investigative and criminal justice reporter at the *Oregonian,* where she began her career. At the *Times,* her work has chronicled the rise and fall of a troubled small-town mayor, discrepancies in the state's counting of coronavirus deaths, and dangerous working conditions inside a Tampa lead factory. The latter earned her and her reporting partners, Corey G. Johnson and Eli Murray, the 2022 Pulitzer Prize for Investigative Reporting, a George Polk Award, and several other national honors. Woolington grew up in Portland and attended Portland Community College before graduating from the University of Oregon. She joined the *Times* in 2018.

KATHERINE J. WU is a staff writer for *The Atlantic,* where she covers science. She's also a senior producer of The Story Collider and a senior editor at the Open Notebook. She previously served as a science reporter for the *New York Times,* where she reported on the COVID-19 pandemic. She won a Science in Society journalism award in 2021 and the Evert Clark/ Seth Payne Award for Young Science Journalists in 2020. She holds a Ph.D. in microbiology from Harvard University.

RAE WYNN-GRANT, PH.D., is a wildlife ecologist, author, storyteller, media host, and advocate. She has focused on the ecology and conservation of large carnivores across the world and is currently a research faculty member at the Bren School of Environmental Science and Management at the University of California, Santa Barbara. She has a passion for storytelling and has recently become fascinated with environmental history, especially as it pertains to the Black experience.

Other Notable Science and Nature Writing of 2021

The Crow Whisperer. *Harper's*,
April 1, 2021.

Rachel May
Mycelium. Guernica, July 22, 2021.

Kat McGowan
Has the Fountain of Youth Been
in Our Blood All Along? Popular
Science, September 28, 2021.

Megan Molteni
Fatal Flaw. *Wired*, May 13, 2021.

Kate Morgan
The Demise and Potential Revival
of the American Chestnut (or, Once
Upon a Tree). *Sierra*, February 25,
2021.

Helen Ouyang
The City Losing Its Children to
HIV. *The New York Times Magazine*,
August 31, 2021.

David Owen
Promised Land. *The New Yorker*,
February 1, 2021.

Mukta Patil
Living with Fire. *Bay Nature*, June
23, 2021.

Daniel Pauley
What Netflix's *Seaspiracy* Gets Wrong
About Fishing, Explained by a Marine
Biologist. Vox, April 13, 2021.

Chanda Prescod-Weinstein
Enter the Axion. *American Scientist*,
May–June 2021.

Marion Renault
A Truly Revolting Treatment Is
Having a Renaissance. *The Atlantic*,
June 2, 2021.

Tim Requarth
Our Worst Idea About Safety. Slate,
November 7, 2021.

Adam Rogers
All in Your Head. *Wired*, December
2021.

Joshua Rothman
Missing a Beat. *The New Yorker*,
March 8, 2021.
Thinking It Through. *The New
Yorker*, August 23, 2021.

Elizabeth Royte
Will This Court Case End the
Mining Industry's 150-Year
Dominance of the West? *Mother
Jones*, August 2, 2021.

Scott Sayare
The Odor of Things. *Harper's*,
November 15, 2021.

Jenna Scatena
Climate Change Is Going to Be
Gross. *The Atlantic*, December 18,
2021.

Zoë Schlanger
What to Save? Climate Change
Forces Brutal Choices at National
Parks. *New York Times*, May 18, 2021.

Sarah Scoles
The Broken Shield. *Scientific
American*, June 2021.

Gary Shteyngart
My Gentile Region. *The New Yorker*,
October 11, 2021.

Jordan Michael Smith
Can a Radical Treatment for
Pedophilia Work Outside of
Germany? Undark magazine, June
7, 2021.

James Somers
Head Space. *The New Yorker*,
December 6, 2021.

Zach St. George
As Climate Warms a Rearrangement
of Plant Life Looms. Yale
Environment 360, June 7, 2021.

Sarah Stillman
The Migrant Workers Who Follow
Climate Disasters. *The New Yorker*,
November 1, 2021.

Jennifer Tsai
Jordan Crowley Would Be in Line
for a Kidney—If He Were Deemed
White Enough. Slate, June 27,
2021.

Debbie Urbanski
Inheritance. *Sun*, June 1, 2021.

Aliya Uteuova
After Slavery, Oystering Offered

EXPLORE THE REST OF THE SERIES!

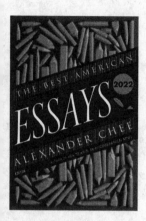

On sale 11/1/22
$17.99

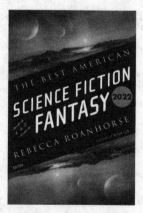

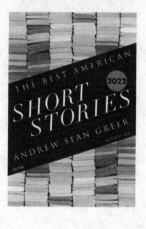